AF559799

WORKSHOP PRACTICE

WORKSHOP PRACTICE

By

Shashi Kant Yadav

DISCOVERY PUBLISHING HOUSE
NEW DELHI-110002

Published by:

Tilak Wasan

DISCOVERY PUBLISHING HOUSE PVT. LTD.
4383/4B, Ansari Road, Darya Ganj
New Delhi-110 002 (India)
Phone : +91-11-23279245, 23253475, 43596065
E-mail : discoverypublishinghouse@gmail.com
sales@discoverypublishinggroup.com
web : www.discoverypublishinggroup.com

***Edition:* 2020**

ISBN: 978-81-8356-107-5

Workshop Practice

Printed at:
Infinity Imaging Systems
Delhi

Preface

Workshop practice has been expanding explosively during the past decade and the initial concept of this book was simply to collate some of the newer and easy applicable methods particularly those involving some degree of automation. This plan was altered in favour of a treatise that not only brings workshop methodology uptodate but also includes representative protocols for the application of these techniques. Accordingly, over a hundred authors have pooled their efforts to produce this volume. In so doing, the mutual hope is that it will serve as a reference portfolio to help both the novice and veteran researcher "get on with the job" in an inspired efficient, and productive manner. On the whole graduating students of most streams of Engineering may find interest in this book and will benefit having one at hand.

In the preparation of this book large number of books and research papers have been consulted. So no authenticity is claimed.

The author wishes to express his deepest appreciation to the many people who have contributed in one way or the other to the preparation of this title.

The author expresses his gratitute to Mr. Wasan and staff of M/s Discovery Publishing House for their whole hearted co-operation in the publication of this book.

The author tried hard to be accurate and upto date in statement and realises the impossibility of completely avoiding errors therefore, the author will greatly appreciate having his attention called to any questionable statement.

Author

CONTENTS

1

General Introduction

MANUFACTURING PROCESS

The production of work pieces having defined geometric shapes is called manufacturing. It is one of the most important production technologies. Process technology (production of chemicals etc.) and energy technology (production of electricity etc.) are other technologies.

CLASSIFICATION OF MANUFACTURING PROCESSES

Manufacturing processes can be classified in six groups. The are:

1. Primary Shaping or Forming Processes

Manufacturing of a solid body from a molten or gaseous state or from an amorphous material is called primary shaping or forming.

Amorphous materials are gases, liquids, powders, fibres, chips, melts and like. To the form of the product a primary shaping or forming tool contains a hollow space, which with the allowance for contraction usually corresponds. Here cohesion is normally created among particles. Some of the important primary shaping processes are:

1. Casting. 2. Powder metallurgy.
3. Plastic technology.

2. Deforming Processes

To cause plastic deformation of the materials to produce required shapes without changing its mass or material composition deforming processes make use of suitable stresses like compression, tension, shear or combined stresses. In forming, no material is removed; they are deformed and displaced. Some of the forming processes are:

1. Forging.
2. Extrusion.
3. Rolling.
4. Sheet metal working.
5. Rotary swaging.
6. Thread rolling.
7. Explosive forming.
8. Electromagnetic forming.

3. Machining/Removing Processes

To generate the surface required by providing suitable relative motions between the workpiece and the tool, this principle is used in all machining processes. In these processes material is removed from the unwanted regions of the input material. In this as compared to forming processes the material is subjected to a lower stress. Some of the machining processes are:

1. Turning.
2. Drilling.
3. Milling.
4. Grinding.
5. EDM.
6. ECM.
7. Shaping and planning.
8. Ultrasonic machining.

4. Joining Processes

To make sub-assembly or final product in two or more pieces of metal parts are united together in this process. The joining process can be carried out by fusing, pressing, rubbing, riveting or any other means of assembling. Some of the important joining processes are:

1. Pressure welding.
2. Diffusion welding.
3. Brazing.
4. Resistance welding.
5. Explosive welding.
6. Soldering.

5. Surface Finishing Processes

To provide intended surface finish on the metal surface of a job these processes are utilized. By imparting a surface finishing process, dimension of the part is not changed functionally; either a very negligible amount of metal is removed from or certain material is added to the surface of the job. Surface cleaning process in also accepted as a surface finishing process. Some of the surface finishing processes are:

1. Plastic coating.
2. Metallic coating.
3. Organic finishes.
4. Inorganic finishes.
5. Anodizing.
6. Buffing.
7. Honing.
8. Tumbling.
9. Electroplating.
10. Lapping.
11. Sanding.

6. Material Properties Modification Processes

In this type of process, material properties of a workpiece is changed in order to achieve desirable characteristics without changing the shape. Some of the processes are:

1. Heat and surface treatment.
2. Annealing.
3. Stress relieving.

TYPES OF PRODUCTION SYSTEMS

It is very important to provide here the matter regarding the production system of an organization because to a large extent, layout, planning and inventory subsystem are directly depends on the production system.

The types of the production can be of three types. These are as follows: (1) job, (2) batch, (3) flow or process production. The simplest way is to classify production proceses by lot size, namely single unit production, small lot production, medium lot production, large lot production, and continuous production.

Whether the production process is continuous or

intermittent, this is determined by the second method of classification. This is apparently a simpler method compared with the first one, but completely ignores lot size or the scope of production.

The categories are included in third classification, these are smallscale production of a large variety of products, medium-scale production of a limited range of products, and a large-scale production of a small variety of products. Obviously, this method is related to the number of product types and production lot sizes, and is an effective means of analyzing modern production management.

The fourth classification is related to the size of the production system expressed through the number of employees or the mount of fixed assets involved, namely small production unit employing less than 20 workers, medium-small having 21 to 300 workers, medium with 301 to 1,000 workers and large with 1,001 to 10,000 people, and a giant corporation employing more than 10,000 people.

PLANT LAYOUT

The arrangement of the physical elements of facilities which results in the production of goods or services is known as plant layout. Thus facilities layout is the overall arrangement of machines, men, materials handling, service facilities, and passage required to facilitate efficient operation of production system.

OBJECTIVES OR GOALS OF PLANT LAYOUT

Like the basic objective of any other organizational facility, the basic objective the plant layout is to maximize the profit of an organization, plant mix-up the basic manufacturing equation in an organisation like men, materials, machines, and money in fulfilling this goal. Objectives of plant layout are given as follow:

1. Minimize materials handling.
2. Reduce the manufacturing cost.
3. Afford economically output quantity and quality.

4. Maintain flexibility of arrangement and of cooperation to meet changing needs.
5. Maintain high turnover of work-in-process.
6. Reduce investment in equipment.
7. Make effective and economical use of floor space.
8. Utilize equipment and facilities in the best possible way.
9. Reduce work delays and stoppages.
10. Promote effective utilization of manpower.
11. Avoid back-tracking.

DIFFERENT TYPES OF LAYOUT

In an organization the facilities and departments are arranged in such a way that they make working easy, departments and facilities arranged may be viewed either in terms of work flow or the function of the productive system. According to *work flow pattern,* four basic types illustrated in Fig. 1.1 and 1.2 are used in manufacturing operations. These are:

1. Fixed-position layout.
2. Product or line layout.
3. Process or functional layout.
4. Group layout.

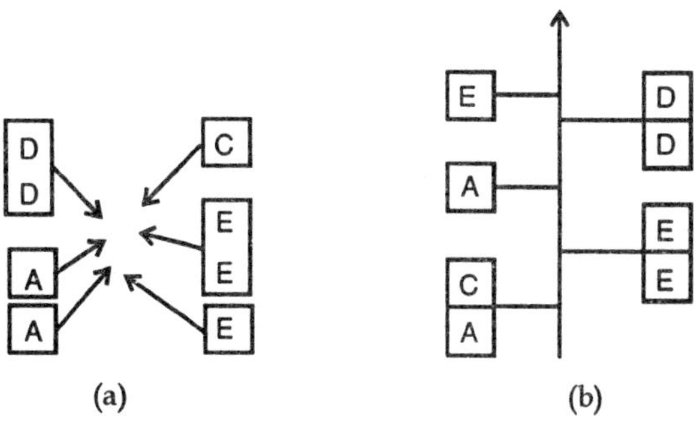

Fig. 1.1 Two types of basic layout are

(a) Fixed layout, (b) Product layout

Group layout is based on group technology philosophy is accepted as the fourth type of layout. In this type of layout an existing functional layout is converted to group layout to derive all benefits of product layout.

1. Fixed-Position Layout

By virtue of its bulk or weight the product remain at on location in the fixed position layout. The equipment, personnel required for the product manufacture are moved to the product, rather vice versa. This layout is observed to be used under following conditions:

1. When material forming or treating operation requires only tools or simple machines.
2. When only one or a few pieces of an item are to be manufactured.
3. When the cost of moving the major piece of material is high.
4. When the skill of workmanship lies in the abilities of workers or when it is designed to fix the responsibility of product quality on one workman or group of workers.

Aircraft, locomotive, sound stages on a movie lot, ship assembly shops are the examples of fixed position layout.

Advantages of fixed position layout are:

1. Handling of major assembly is reduced.
2. Highly skilled operators are allowed to complete their work at one point and responsibility is fixed on one person or assembly crew.
3. Instant changes in product design and in sequence of operations are possible.
4. The arrangement is adopted to variety of products and intermittent demand.
5. It is more flexible almost in all position in plant.

2. Product Layout

When in a plant the assembly work stations and equipment are arranged according to the progressive steps of sequence by which the product is manufactured is called product layout. Product layout is used advantageously in the continuous production system where:

1. The number of end products is small.
2. The parts are highly standardized and interchangeable.
3. The volume of production for any item is high or large.
4. The demand of the product is fairly steady.
5. The continuity of material flow can be maintained.

This type of layout is exemplified in automobile assembly, food processing, and furniture manufacture.

Advantages of product layout are as follow:

1. Reduced handling of material by transfer machines or conveyors.
2. Reduced amount of material in process, allowing reduced production time, and lower investment in materials.
3. More effective use of labour through:
   (i) Greater job specialization.
   (ii) Ease of training.
   (iii) Wider labour supply.
4. Easier control.
5. Less inspection, less work-in-process, less floor area, and smaller aisles.

Disadvantages of product layout are given below:

1. Higher initial equipment investment.

2. Greater vulnerability to work stoppage.
3. Greater inflexibility due to changes in design, making it more costly.
4. Greater boredom due to highly repetitive nature of work.

3. Process Layout

When all operations of the same type are arranged in group according to the general function perform by them with regard to any particular product than it will called process layout.

Conceptually, same type of operations are performed in the same area. For instance, all drilling machines may be grouped together in one area, all milling equipment will be in another work center, and work involving spot welding is still another, etc. Refer Fig. 1.2.

The process layout is used better where:

1. Volume is low.
2. Variety of products is many.
3. The demand for the product is small or intermittent.
4. General-purpose, low-cost machines are desirable.
5. Machines or processing producing unwanted noise, vibration fumes, or heat can be located in isolated areas.

This type of layout is exemplified in custom jobshop, department stores, hospitals, etc.

Advantages of this type of layout are given under:

1. Better machine utilization allows lower machine investment.
2. Adaptable to a variety of products and to frequent changes in sequence of operations.
3. Readily adaptable to intermittent demands.
4. Makes easier the continuity of product in the event of:

(i) Machine break-down.

(ii) Storage of materials.

(iii) Absent of workers.

5. Permits easier maintenance.
6. Allows incentives pay system to individual workers to raise the level of their performance.

Disadvantages of process layout are given under:

1. Operations cost per unit may be higher—often much higher because of the slow operation of the general purpose machines used.
2. Work routing, scheduling, and cost of accounting are costly due to separate costing of every new order.
3. Materials handling and transportation costs are high since products follow different routes every time and use of conveyor becomes uneconomical.
4. Inventories of materials in process are high and large storage space is required because of slow movement of materials.
5. Difficult to keep a good balance between labour and equipment needs.

4. Group Layout

In group lay out the group technology is followed. In this process the every batch or shop is divided in the different groups of machines. In this, the each group produced its associated families of components. Group technology (GT) philosophy uses similarity of components from the processing point of view to create component families. By using the philosophy, a job-shop is transformed into a group technology environment which facilitates optimization of production scheduling.

The essence of GT makes use on similarities in recurring tasks in many ways, like

1. By performing similar activities together thereby avoiding wasteful time in changing from one unrelated activity to the next.
2. By standardizing closely related activities thereby focusing only on distinct differences and avoiding unnecessary duplication of effort.
3. By efficiently storing and retrieving information related to recurring problems thereby reducing the search time for the information and eliminating the need to solve the problem again.

Fig. 1.2 is showing how a job shop can be partitioned to GT shops. It is noted that the four components are the representative components of the component families.

Advantages of the group layout are as follow:

1. Reduction in machine set-up time and cost.
2. Reduction in material handling cost.
3. Elimination of excess work-in-process inventory which subsequently allows the reduction in lot size.
4. Reduction in through-put time.
5. Simplification of production planning functions, etc.

Disadvantages of the group layout are as follow:

1. Change of the existing layout to GT layout is time consuming and costly.
2. Inclusion of new components in the existing component families requires thorough analysis.
3. Change of input component-mix may likely to change cell structure.
4. Change in batch-sizes may change number of machines for any machine type.
5. Existence of a large number of exceptional components leads to complexity in machine loading.

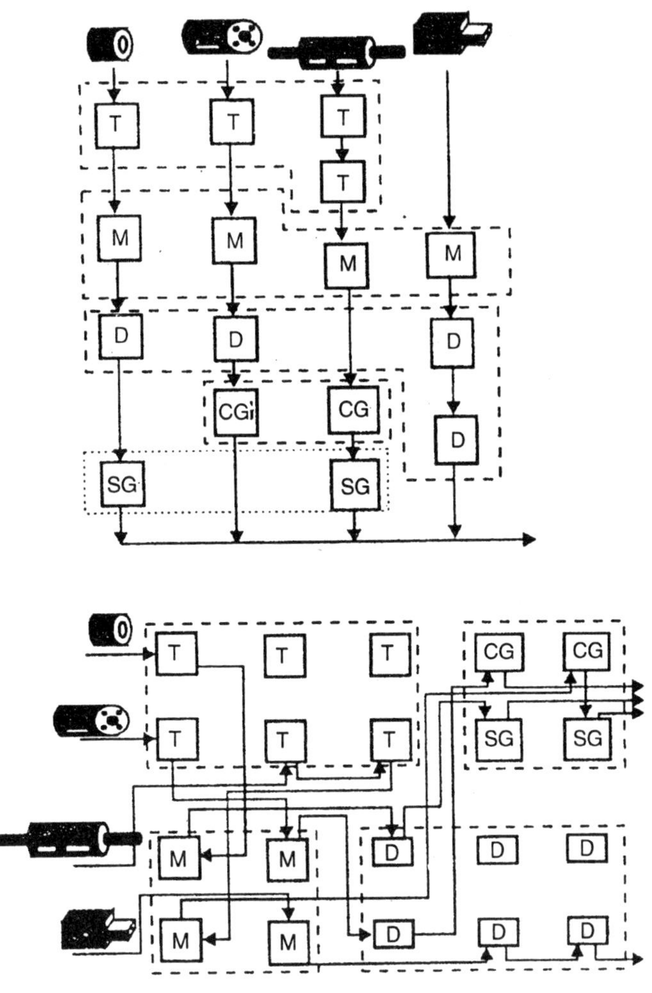

Fig. 1.2 Types of layout

(a) Basic type : process layout (b) Group layout.

MANUFACTURING AND BASIC DEFINITIONS

Manufacturing can be defined as the collection of interrelated activities, includes design of product, selection of materials, documentation, planning, quality production, management of activities and marketing of goods or product.

The objective of manufacturing activities is to convert raw materials into finished goods on a profitable basis.

Productivity: The basic meaning of producitivity is "the quality or state of being productive". This word has been originated from the understanding of the meaning of productive efficiency, which is largely responsible for the growth of the economic condition of any industry or industries as a whole resulting in the prosperity of the society for improving the standard of living of its people. Productivity is the output-input ratio within a time period with due consideration for quality. In short,

Productivity = output/input or P= O/I

This concept of productivity denoted the effectiveness and efficiency of individual and organisational performance.

Different techniques of productivity improvement may be classified into five basic types. These are as follow:

1. Employee-based techniques
2. Task-based techniques
3. Technology-based techniques
4. Material-based techniques
5. Product-based techniques

Interchangeability: If any part from a lot will fit to its counterpart than it will known as "interchangeability". This can be achieved by maintaining tolerances for part dimensions that takes into account the manufacturing tolerance of mating parts. The result of design for interchangeable parts is interchangeable assembly. Assembly costs then become much lower than they would be if workers had to select combinations of mating parts that fit.

Just-in-time (JIT) Management: The other many names of JIT are stockeless manufacturing, lean production or synchronous manufacturing. JIT is a Japanese management philosophy. It created the dominent importance in the mind of manufacturing planners. The JIT philosophy evolves from three principles: reduction of production cost, elimination of waste and the recognizing workers abilities. The JIT strives to reduce *work in process* (WIP) by producing only the minimum number of required parts. It has been described as the *demand pull* rather than a *schedule push*. This description is based on the pulling action of KANBAN (a kind of *visible record* or *card)*, the production control system used to ensure that the right amount of parts is made at the right time. In a JIT system, parts or components are not produced until these are needed by the downstream work-centres. The sub assembly and parts requirements are actually determined on the final assembly line and passed upstream to supply work centres. This pull system works because of the commitment to many other elements that make up the JIT philosophy.

AUTOMATION, MECHANIZATION, CAD, CAM AND CIM

Automation and Mechanization: It is a kind of set up, where human functions or attributes are replaced by the machine or group of machines, on other hand mechanization is the preliminary atomacity, which includes semiautomatic machines.

Automation in manufacturing can be classified into two forms:

1. Fixed position automation, and
2. Soft automation.

By the equipment configuration the sequence of processing operations is fixed than it will called fixed automation. This type of automation is achieved by cams, gears, wiring and other hardware. This configuration cannot be easily changed over quickly from one product types to other.

High production rates and high initial investment is the prerequisites of this type of system. Transfer line/machines, automatic screw machines, automatic assembly lines are the few examples.

Soft automation can again be divided in two types:

programmable and *flexible.* In programmable automation, the production system is capable of manufacturing a variety of product types through numerical programming. However, changeover time, from manufacturing one type to the other type of component, cannot be eliminated. A numerical control machine is the example of this type. Generally a batch of components is produced at a time till the changeover takes place. Production rates in programmable automation are generally lower than that of fixed automations.

In flexible automation, the changeover of the equipment for manufacturing different component is very quick and automatic. The programming of the machine/facility is done off-line. In this type a mixture of different products can be produced one after another without any delay in between. One example of flexible automation is flexible machining system/centre.

Advantages of automation as follows:

1. It increased the productivity due to high throughput.
2. It lower the labour costs.
3. It is safe in processing.
4. It provide efficient utilisation of materials.
5. It provide reduction of in-process inventory.

Computer-Aided Design (CAD). CAD is the process by which we use the computer hardware and graphics software to enhance the product design from conceptualization to documentation. In utilizing CAD, the computer draws pictures on the video screen or computer controlled cathode ray tube (CRT) of what the engineer has designed. The sophisticated CAD software is able to draw lines between points, creating a three-dimensional drawing that can be rotated to show all sides of the design.

Computer aided Manufacturing (CAM) can be defined as the implementation of a wide range of automation technologies. It is the effective and efficient use of computer hardware and software in planning, management and control of production of an organisation.

Computer-Integrated Manufacturing (CIM) is the combination of all aspects of productive enterprise through the use of integrated systems and data communications coupled with new managerial philosophies that improve organizational and personal efficiencies and effectiveness. In a CIM, flexible automation of all manufacturing activities is realized. It represents the direction of the development of modern manufacturing.

Following are the advantages of CIM:

It improved customer service.

It improved quality.

It reduce inventory level.

It lower total costs.

It increase productivity.

It lower in WIP inventory.

It high through-put.

INDUSTRIAL SAFETY—THE PRIME OBJECTIVE OF A CONCERN

Occasional accident may occur in an industrial unit but prime objective of a concern is to save the men, material and power from the accidents. Accidents results in the direct and indirect losses to the plant.

Accident—Causes

A large number of accidents can be avoided if proper safety rules are followed. Under given are the basic causes of accidents in the plant:

1. Violation of safety rules and not using safety devices.
2. Improper use of gadgets and machine controls.
3. Ignorance of the system operation.

4. Unsafe working conditions.
5. Monotony and work-relating stresses.
6. Wear and tear of the functional components.
7. Organic diseases etc.

Two basic emotional factors that have been found to be related to accidents in many of the work situations are general emotional maturity and emotional conditions of the workers at the time of accidents. Mental ability is an important personal factor that has been co-related to accident frequency. For some jobs a certain minimum mental ability level is required if the operator is to escape dangers that is generated from the operation.

Another factor that is co-related to the accident-proneness of an individual is his lower perceptual and higher muscular speed.

Accident—some preventive measures

The following safety measures are only a few of the otherwise long list:

1. Use of the proper safety guards for reciprocating machine components (e.g. drop hammers, paper cutters, etc.).
2. Fencing of dangerous and rotating parts like revolving shafts.
3. Incorporating safety devices (safety valves/ facilities.)
4. Rigid construction of heavy items like hoists, cranes, etc.
5. Proper insulation of electric wire and earthing of electric appliances.
6. Wearing appropriate safety shoe and other necessary items for body protection.
7. Maintenance cleanliness of shop floor.

8. Removal of metal chips with proper protection.
9. Avoiding fire hazard.

Electrical Safety

Some rules and regulations have been formed on the basis of plant experience, while handling the electric appliances. If a person receives electric shock, mild or serious, directly or indirectly, the same must be reported as an electrical accident. In industry, an inquiry is normally ordered for every electrical accident. The following table shows the effect of electric current on human system.

TABLE 1.1 : EFFECT OF ELECTRIC CURRENT ON HUMAN SYSTEM

Electric Current	*Effect*
1 to 8 ma	Perceptible but not painful
8 to 15 ma	Painful shock; muscular control is not lost
15 to 20 ma	Painful shock; muscular control is lost
20 to 50 ma	Severe muscular contraction; breathing difficult
50 to 100 ma	Ventricular fibrillation; a heart condition that results in instant death.
200 ma and above	Severe burns, clamp down the heart and stop it during the shock. This prevents the ventricular fibrillation.

All appliances carrying electricity/having electric potential will have to be properly insulated. As the insulation have specific dielectric strength or voltage withstand capacity, the working voltage must, therefore, be kept substantially below the breakdown value. Ageing of insulation reduces the withstand capacity and should not be allowed to leak current. Switching contacts and point contacts should be checked periodically to avoid sparking and point heating. Under given are some preventive measure, one must follow:

1. All metallic parts, externally accessible must be earthed.
2. Defective and worn-out fittings must be replaced promptly.

3. Inflammable materials should not be kept near electric appliances, live wires or electric control panels.
4. Safety devices like fuse, circuit breakers over-tripping switches must be used when required.
5. Untrained persons should not be allowed to repair electrical faults.
6. Safety belts, helmet, rubber gloves, boots with rubber soles, and other safety items must be supplied to electric mechanics.
7. Standards of safety items must be maintained as per norms.
8. Avoid overloading any electric circuit.
9. Disconnect the electric appliances from main supply when not in use.
10. Avoid touching metal case of any electrical apparatus when it is in operation.

Fire Safety

Fire accidents are occur during the working period in an organization which caused damage to man, property and materials. Fire accidents can be classified in several ways. One such classification is shown in Table 1.2.

It can be noted that the origin of fire is combustion. The three factors for combustions are:

1. Presence of oxygen.
2. Availability of combustible materials.
3. Rise of temperature to the ignition point for the material.

Following are some causes of fire accident in organisation:

1. Electric-short circuit.

2. Carelessness and ignorance.
3. Spontaneous combustion.
4. Riots, arsons, rivalry, etc.

Preventive measures for preventing fire accident:

1. Preventing speading of fire by covering it with incombustible material.
2. Rescuing the affected persons from the spot.
3. Cooling ignition point of combustible items.
4. Taking suitable precautions and educating the workers about the possible hazards out of the consumable that spread fires.
5. Color coding of pressure vessel containing combustible fluids.
6. Maintaining fire extinguisher and water points in shop floor for proper precautions in case of fire hazards.

Accidents And The Factories Act

These are various acts regarding accident compensation. i.e. Workmen's Compensation Act, 1923, The Factories Act, 1948, and Fatal Accidents Act, 1855.

Workmen's Compensation Act,1923. *Section 3* of Workmen's' Compensation Act provides employer's liability for compensation. Compensation is dependent on four conditions:

1. Injury must be personal.
2. Injury must be by an accident.
3. Accident must have arisen out of and during employment.
4. Injury must have resulted either in death or total or a partial disablement for a period exceeding three days.

TABLE 1.2 : CLASSIFICATION OF FIRE ACCIDENTS

Accidents			
Indoor	**Outdoors**	**Basic causes**	**Organizational defects**
• Petrochemical & Fertilizer plants • Manufacture of explosives & pyrotechnics • Factories and workshops • Construction Premises • Hospitals and school • Hostel and clubs • Offices and shops • Warehouses and storage depots • Homes and dwellings • Wood structures, thatched huts • Temporary pandals & shamianas	• Coal dumps • Oil and gas storage tanks and pipe lines • Outdoor machinery & equipment • Road trucks and cars • Ship & Aircraft • Railway coaches • Coal mines • Grasslands and forests • Timber depots • Ignition of hot gases • Lighting	• Arson • Careless throwing of cigarettes, matches • Poor housekeeping & dumping of trash • Spontaneous combustion • Electric sparks overheating welding • Ignorance and forgetfulness • Gas & oil stoves in the kitchen • Candles and oil lamps • Playfulness of children and fire-crackers • Wearing nylon sarees in kitchen	• Inadequate fire-fighting facility • Poor communication and transport facilities • Absence of fire-drills and poor training of Staff • Poor design of buildings; lack off exits & fire-escapes • Lack of fire detection equipment • Improper and poor maintenance • Delayed first aid & medical help

Section 4 lays down the amount of compensation payable to the workers for four categories ranging from temporary disablement to death. *Section 10* also specifies the employer's liability to report fatal accidents to an authority.

Factories Act, 1948. This act defines various terms of factories like manufacturing process, machinery, worker, power, prime movers etc. *Section 21* lays down rules for fencing of machinery for safeguard and also the provision for work on or near machinery in motion. *Section 23* of the act provides clauses to ban employment of young persons on dangerous machines.

According to Factories Act, 1948 various machines are considered as dangerous, these are given below:

1. Decorticator and expeller.
2. Milling machines used in metal trades.
3. Guillotine machines.
4. Power presses, other than hydraulic presses.
5. Planten printing machine.
6. Circular saw.

Section 24 specifies provision for striking gear and devices for cutting off power for safety. Other measures and provisions are self acting machines (*Section 25*), Hoist and lifts (*Section 28* by Act 20 of 1987), Revolving machinery (*Section 30*), Pressure plant (*Section 31*), Precaution against dangerous fumes (*Section 36*), and Precaution in case of fire (*Section 38*). Section 88 places an obligation on the management to notify accidents.

Fatal Accidents Act, 1855. Two kinds of damages are provided by this act. These are:

1. Damages proportionate to the loss caused by the death to the beneficiaries, and
2. Damages to any pecuniary loss to the estate of the deceased.

2

Material Testing—Treatment and Properties

We all know that our ancestor basically the orthodox type simple metals and aloys, which were used by them. The quality and variety of metals they possess were very limited. On other hand these days we possess the almost endless variety of metals and we can use them in different manner in various type of works for a specific type of job. There is now a specific metal. This has been achieved by advances that have taken place in metallurgical science and research working in conjunction with manufacturing processes and industrial needs. The forms in which these materials may be obtained is also a tribute to our industrial skill, and it would be an unreasonable critic who complained of inefficiency on the part of our material suppliers.

The knowledge of properties and application of the metals used in the workshop engineering essential for the students. He should have a fair idea of their cost also, as the economic factor is one which should never be overlooked. Later in this chapter we provide discussion of some of the additional metals which are being developed, and we advise the students to supplement his knowledge regularly from current research and periodicals, since progress in this field is constantly taking place.

The cost of materials is not a proper subject upon which to give information in a textbook as costs vary from time to time. For this purpose the students should make a habit of glancing through the metal-market details in the weekly journals dealing with engineering and the metal trades. A knowledge of the varieties of shapes and forms in which the chief metals may be obtained is very crucial as such knowledge, well applied, may save much time and expense in the production of an article. Perhaps, by using the material with certain of its shapes or dimensions accurately finished in the primary rolling or drawing process, expensive machining operations may be omitted, and a saving in cost achieved. Our application of this knowledge is made more effective if we know the possibilities of accuracy and finish, and the cost associated with various forms of supply.

Now here we are discussing the example of steel strip. The can be obtain in following forms:

(a) Hot rolled. This type of steel strip may find in reddish, and scaly surface. It need polishing or machining for some specific purposes. The thickness will probably be finished to within ± 0.1 mm of the nominal size and the width to within ± 0.5 mm on the narrow sizes, increasing slightly on wider strip. The edges of this steel is generally found in slightly rounded. It is not flat and after cooling from rolling temperature the material can be find.

(b) Cold rolled. The general characteristics of this steel are bright and smooth surface which need not additional finishing process or polishing. The thickness can be controlled to within about ± 0.01mm, but of course the more stringent the limits imposed, the higher the cost. The edges of this strip will be slightly rounded and the width (to about ± 0.5mm) will be measured to the corners where the rounded portion starts. For a slight additional cost this strip can be supplied sheared to width. It will then have square edges and width limits of the order of ±0.1 mm. If the strip is supplied 'as rolled' it will be in the workhardered condition, springy and liable to crack if bent. These wokhardening effects can be

removed without impairing the bright surface of the strip by subjecting it to a special close-annealing process which leaves the strip with its bright surface but brings the metal to its normal condition by removing the effects of workhardening.

There are various methods by which close annealing can be carried out, but the object in very case is to prevent oxygen from acting on the surface of the strip while it is at the annealing temperature. In one method the strip is loaded into the furnace packed in iron or steel boxes along with cast-iron swarf, while in a more modern method the atmosphere inside the furnace is rendered non oxidizing by maintaining a circulation of a gas such as cracked town gas, which not only prevents oxygen from entering, but also has an active effect in preventing the surface of the strip from oxidizing. Naturally, because of the additional processes to which it must be subjected, bright strip is more expensive than the hot-rolled (black) variety.

SUPPLY FORMS

To know regarding the various forms of supply of the materials is very important point for the students, because the physical properties and other characteristics of the materials are very different. Generally these factors impose limitations in the form of supply of material or about the use of a metal for specific purposes. For instance, cast iron in its solid state is brittle at all temperatures and for this reason can be neither rolled nor drawn. If, therefore, we require cast-iron bar, we can obtain only short bars because it has tobe cast into this form and the limitations of this process render the production of long bars almost impossible. Even if long bars could be produced, it would be extremely difficult to handle and transport them without breaking! Again, while steel may be melted and cast, it never attains a fluidity comparable with that of many other metals and alloys. For this reason, steel castings cannot be relied upon for perfection where intricate shapes and clean surfaces are an important consideration or where reasonably close and uniform size limits must be maintained; such qualities are more easily achieved in drop forgings.

In Table 1 is showing the chief forms in which metals are available for use in the workshop.

SPECIFICATION OF MATERIALS

Specification of the material is a process by which the quality and composition of the materials are set out together satisfy in order to be suitable for its purpose. In addition, the specification will lay down any other conditions which may be necessary to control the supply of the material to which it refers.

There are so many advantages of specifications. It promote a feeling of confidence between the contractual parties and produce possible and satisfactory outcome. It will favour the efficincy of supply of material and industry, as well as the public, is assured of a reliable standard of quality in the commodities it receives. If manufacturing firms work to a limited number of specifications having national standing, they are enabled to concentrate on these products, thus avoiding the waste resulting from working to small, special orders.

British Standards Institution (BSI), which works in close collaboration with the International Standards Organisation (ISO), particularly as regards metrication. Under its auspices are issued specifications covering every aspect of industry, and it would be difficult to estimate the benefits that have resulted from its work. We reproduce in Appendix 1 the main portion of one of its specifications – BS3100: 1967, 'Specification for steel castings for general engineering purposes,

In the process of specification certain physical conditions and certain tests must be passed by the materials.

In the example we have shown, clause 1,11 (mechanical tests) specifies certain testing procedures for steel castings. It will be well, therefore, for us to give some attention to the more common tests to which materials must be subjected.

Table 2.1. Forms of supply of chief workshop metals

Metal	Chief forms of supply	Remarks
Cast iorn	*Standard forms and those obtainable without much difficulty*	
	Short bars, hollw cylinders, blanks, cast plates, pipes.	Samll pipe fittings and small castings, e.g. cheap builder's hardware, etc., often of malleable iron.
	Pipe fittings. Miscellaneous standard castings, e.g. bearing brackets, pulleys, firebars pots, legs, etc.	Large pipes, often spun cast. Surface of castings may be rendered hard by 'chilling'.
	To special order	
	Castings to pattern.	
Low-and mediium-carbon steels	*Standard and common forms*	
	Hot rolled (black) bars; round square, flat, etc. billets.	Structure of black steel is that given by air cooling from rolling temperature.
	Rolled sections, T, L, I, etc.	See BS specifications for rolled sections.
	Black sheets, plant and strip, black tubing.	Close limits of accuracy possible in bright-drawn and rolled sections. These processes work-harden the metal which should be 'bright annealed' if normal structure is desired.
	Bright-drawn bars; round, square, hexagonal, rectangular, etc, wire.	
	Bright-rolled flats and strip, drawn tubing.	
	Miscellaneous black and brightforged articles, e.g., rivets, bolts, washers, split pins, nails, brackets, pipe fittings etc. blanks.	Thin low-carbon steel sheet generally supplied with
	'Half and half' in flat bars. (This has a base of	

continued...

Metal	Chief forms of supply	Remarks
	mild steel with a welded layer of medium- or high-carbon steel. Used for dies.)	tinned surface. Nominal size of gas pipe is the bore. Drawn tubing is specified by the outside diameter.
	Corrugated iron (galvanised for weather protection)	
	To special order	
	Black or bright bars, strip, etc. having special dimensions or shape.	Simple forgings and small quantities made by hand or power hammer. Large quantities made by dropstamp or forging machine. Pressings made from sheet or plate by pressing process.
	Forgings of common shapes, e.g., blanks, blocks, stepped shafts, etc.	
	Drop forgings. Hand forgings. Castings. Pressings.	Medium-carbon steel which will harden when quenched is sometimes called 'common hardening steel.'
	Miscellaneous special forms, e.g., expanded metal, perforated sheet, chequered plate, etc.	
High carbon steel	*Standard and common forms*	
	Black rolled bars of round, square, flat, octagonal, etc. section.	Octagon bar used for cold chisels.
	Black strip (spring steel) spring wire. Short bars approx. 300mm long, diameters ranging up to about 12mm, ground to fine limits on the steel diameter.	Generally known as silver steel.
	Flat strip and plate surface ground to close limits of thickness.	Often called 'ground flat stock'.

Continued...

Metal	Chief forms of supply	Remarks
	To special order Forgings for blanks, die blocks, tools, etc.	
Brass	*Standard and common forms* Bars: round, square, flat, hexagonal, etc. Sheet, strip, tubing, wire. Ingots (for melting down), Cast sticks. Powder (spelter) and rod for brazing. Rivets, eyelets, screws, washers, etc. Miscellaneous standard water fittings.	Cold-drawn and rolled sections may be obtained in the springy workhardened condition ('hard drawn' or 'hardrolled'), in the 'half hard' state or in the natural annealed form. Water fittings often chromplated.
	To special order Castings and forgings (including hotbrass pressings). Cold pressings. Drawn, rolled and extruded bars having special cross-sectional shapes.	
Bronze gunmetal	*Standard forms* Rolled and drawn bars. Short cast sticks (solid and hollow). Rolled stip and sheet. Wire. Ingots.	Hollow sticks useful for making bushes. Hard-rolled strip and wire used for springs.

Continued...

Metal	Chief forms of supply	Remarks
	To special order Castings. Bars with special cross-sections.	
Aluminium alloys	*Standard forms* Rolled and drawn bars. Cast sticks. Rolled sheet and strip. Wire. Tubing. Rivets, washers, etc. Ingots.	
	To special order Sand castings (sand mould). Die castings (metal mould). Forgings. Pressings. Bars of special cross-section.	Close limits and uniformity possible with die castings.

The tests themselves are also covered by BS specifications to which the reader may refer for more detailed information.

THE TEST OF TENSILE

The objective of this test is to know the strength of a material in tension. The behavior of the material during the test is also used as a guide to its ductility or softness. The tensile test is carried out in a *testing machine* on a *specimen* previously made from the material to be tested. The dimensions and form of the specimen vary according to the size and shape of the material to be tested; flat test pieces being used from strip or flat bars and round specimens from round-section bars, etc. The British Standard recommendations for standard proportional test pieces are shown in Fig. 1 and the accompanying table. The gauge length bears a fixed relationship to the test area, i.e. gauge length =5.65 $\sqrt{S_0}$. The reduced centre portion is the test part proper, the ends being for the purpose of gripping in the machine. The extremities of the gauge length are marked with centre-punch dots so that, when a test is completed and the specimen stretched and broken, the broken ends may be fitted together and the stretch (elongation) measured. The test diameter is joined to the enlarged ends by a smooth radius because, if the corners were sharp, a concentration of stress might occur at this point of sudden change in diameter, and the fracture would take place there instead of somewhere between the gauge-length points. This occurrence would prevent any accurate reading of the elongation and might also give a false reading for the breaking load.

The testing machine provide two types of facilities one is to gripping the ends of the specimen and shifting tension on it and other is to compress the end of the specimen whenever the compressive test is essential. Incorporated in the machine are means for measuring and indicating the tensile or compressive force exerted. There are different types of machine which, according to their design, may cause the force to be exerted by hand, mechanical or electrical power, or by hydraulic means. The indication of the force may be

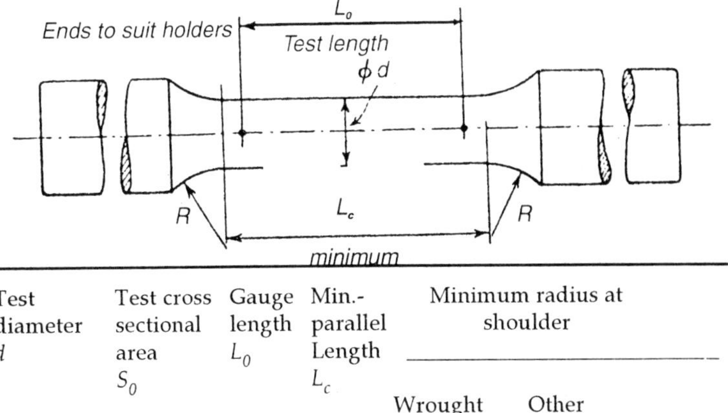

Test diameter d	Test cross sectional area S_0	Gauge length L_0	Min.- parallel Length L_c	Minimum radius at shoulder		
				Wrought metals and cast steel	Other cast metals	Tolerance on diameter
(mm)	(mm^2)	(mm)	(mm)	(mm)	(mm)	(± mm)
25.23	500	125	140	25	50	0.15
22.6	400	113	124	23.5	47	0.125
20	314	100	110	22	44	0.10
16	200	80	88	15	30	0.08
14	154	70	77	12	25	0.07
11.3	100	56.5	62	10	20	0.055
10	78.5	50	55	9	18	0.05
8	50	40	44	7.5	15	0.04
7	38.5	35	38.5	6	13	0.035
5.64	25	28.2	31	5	12	0.030
5	19.6	25	27.5	5	11	0.025
3	7.07	15	16.5	4	8	0.015

Fig. 2.1 : British Standard test piece

effected by some means similar to a weighing machine, or by a pointer on a scale operated by hydraulic pressure of by a diaphragm.

A 200 kN hydraulic machine is consists the two units. indicating unit on the right side and the straining unit on the left side. On right side it contains the pumping and valve gear for the controls and mechanism for indicating the load. The straining unit on the left has a hydraulic cylinder in its base which, when operated, draws down the central cross-head, so that a specimen held between this and the upper cross piece in loaded in tension, while a specimen placed between the cross-head and the planten may be compressed. The two units are connected by the hydraulic pipes and electric cable necessary for operation. Different accessories may be attached to the machine which make possible the tests such as the transverse loading of beams, shear tests, bend tests and timber tests to be carried out. In addition, an automatic stress-strain recorder may be fitted which automatically draws a graph showing the behaviour of the material, similar to those shown in Fig. 2. The load-indicating pointer is operated by a mechanism incorporating a system of levers actuated by a proportional ram energised by fluid from the main loading ram. When making a tensile test, the ends of the specimen are clamped in grips carried in the cross-heads and in the upper cross-beams.

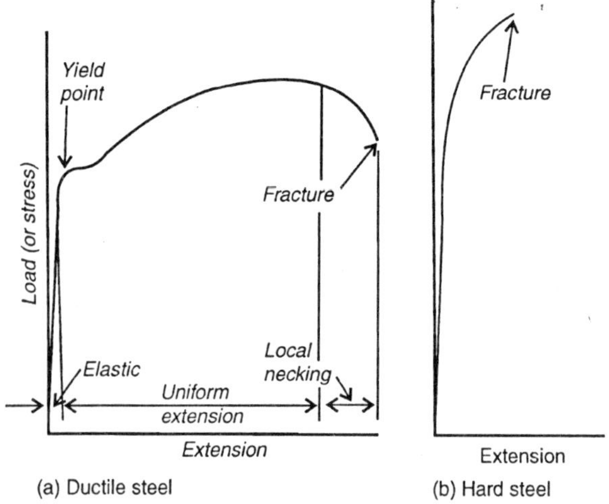

Fig. 2.2: Load (or stress)-extension graphs.

Materials behaves is an elastic manner with a uniform and steady rate of stretch for the increasing load in the early stages of the loading. For ductile steels, a point is reached when this condition terminates and the material stretches as though the material had lost its power to sustain the load. This is known as *yield point.* The material eventually recovers again and more load may by added, although the rate of stretching increases.

Towards the end of the test the specimen will be observed to thin at one point and it may be necessary at this stage to reduce the loading slightly in order to balance the machine. At last, fracture will occur somewhere in the length of the reduced test portion of the specimen. Harder and more brittle materials may not exhibit any pronounced yield point or necking, the specimen suddenly fracturing without any previous warning. If the relationship between load and extension is plotted on a graph, the appearance of such a graph for ductile and hard steels is as shown in Fig. 2.

The results basically required from a industrial tensile test are as given under:

(a) Tensile strength

By dividing the maximum load by the original cross-sectional area of the specimen we can found the tensile strength.

(b) Elongation%

$$\frac{\text{stretch}}{\text{original length}} \times 100$$

by aplying above given formula elongation % can be found. Good elongation combined with a reasonable strength indicates a ductile material.

Besides these two main properties, a specification may stipulate other requirements which may be obtained from the tensile test:

(c) Reduction in area%

This can be obtained by

$$= \frac{\text{reduction in area at fracture}}{\text{original area}} \times 100$$

$$\frac{\text{original area} - \text{area at fracture}}{\text{original area}} \times 100$$

A high reduction of area shows that the material will lend itself more readily to cold working.

(*d*) *Stress at yield point* (when this is present)

$$= \frac{\text{load at yield point}}{\text{cross = sectional area}}$$

Note this point that a stress will be expressed as a load per unit area. For example, if a load in newtons is divided by an area in square millimetres, the units of stress will be newton per square millimetre (the plurals are generally omitted when writing units).

Example. Determine the results of a test on a BS test piece, 14 mm test diameter, 70 mm gauge length, which gave the following results when tested:

Load at yield point.. 51 kN

Maximum load .... 82 kN

Length between gauge point with broken ends put together .... 90 mm

Diameter at fracture. .... 11.5 mm

(*a*) Tensile strength=$82 \times 10^3/154=533$ N/mm^2 (since the area of this test piece is 154 mm^2)

(*b*) Stress at yield point= $51 \times 10^3/154=331$N/mm^2

(*c*) Elongation $\% = \frac{90-70}{70} \times 100 = \frac{200}{7} = 28.6\%$

(*d*) Reduction in area%

$$= \frac{154 - \pi(11.5)^2/4}{154} \times 100 = \frac{154 - 104}{154} \times 100 = 32.4\%$$

Proof Stress of the Metals

A definite yield point do not exhibit by the hard steels and non-ferrous metals when they pulled in the testing machine, for such cases a better measure of their elastic properties is defined by the *proof stress.* This is the stress essential to cause a certain permanent elongation to the test piece (generally about 0.1%), and the method of determining it is shown in Fig. 3.

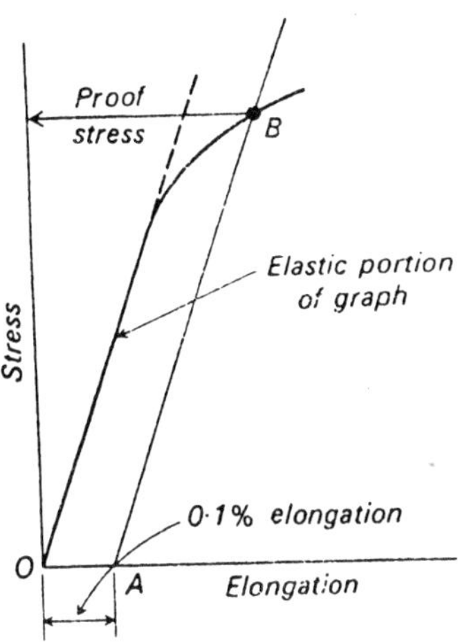

Fig. 3: Proof Stress

A point A is determined, where OA represents the percentage elongation, and then form A a line AB is drawn parallel to the elastic line of the stress-strain diagram. The point B where this line cuts the curve gives the proof stress.

TESTING OF HARDNESS

The Brinell Test

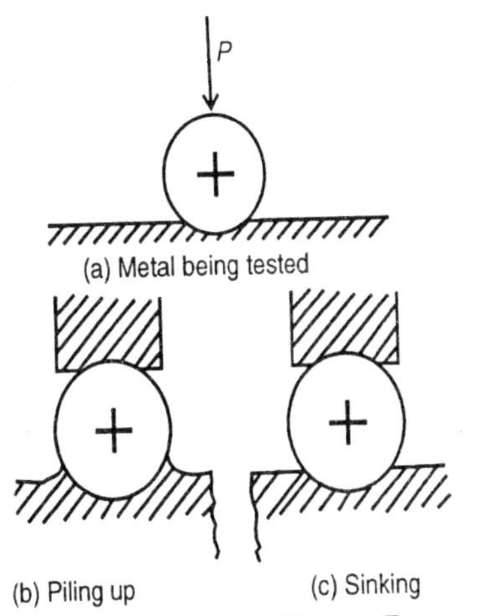

Fig. 1.4 : The Brinell test – Forms of impression

Brinell test is the most well=known test for hardness of metals. A hardened steel ball is pressed onto the metal and an indentation is made, the size of which varies according to the hardness of the material. In Fig. 4 the Brinell test is shown. The Birnell hardness number is found by dividing the *surface area* of the impression (in square millimetres) into the load (in kilogram force). (Note that the surface area is the area of a portion of a sphere.) For the sake of uniformity in results with this test, the relationship between the load and the ball diameter has to be fixed. For testing the hardness of steel, this relationship is given by

$$\frac{P}{D^2} = 30$$

where P=load in kgf and D= ball diameter in mm.

For a 10 mm ball (as is most commonly used)

$$D = 10 \text{ and } D^2 = 100$$

$$\frac{P}{D^2} = \frac{P}{100} = 30$$

from which

$$P = 30 \times 100 = 3000 \text{ kgf}$$

For thin and fragile work a 1 mm ball is often used. Then, from the above relationship, P=30 kgf.

The hardness number which, as we have stated above, is given by load/surface area of impression becomes

$$\frac{3000}{\text{area of impression}}$$

when the 10 mm ball and 3000 kgf load are used.

The surface area of the impression can be calculated as the area of a segment of a sphere when the diameter of the impression has been found. This is usually measured with a graduated microscope.

If d =diameter of impression (mm) and D= diameter of ball (Fig. 4), then curved surface=$(\pi\ D/2)\ [D - \sqrt{[D^2 - d^2)}]$, which; when D=10, becomes $15.71[10 - \sqrt{(100 - d^2)}]$, and

$$\text{Brinell number} = \frac{3000}{15.71[10 - \sqrt{(100 - d^2)}]}$$

Actually it is not generally necessary to calculate Brinell numbers in this way, as charts are available form which the number may be read direct when the diameter of the impression (d) is known.

When softer metals such as copper and aluminium are being tested the relationship P/D^2 =30 would cause the ball to be loaded too heavily and would cause an unduly large impression. For this reason the relationship used for copper is P/D^2=10, and for soft aluminium P/D^2=5.

When making a Birnell test, the surface to be tested should be smoothed and the specimen placed on the table of the machine. The ball should be brought down until contact is made and then the load applied. This should be maintained for a full 15 seconds, after which it may be released and the impression measured with the microscope or whatever means are available. The hardness number may be calculated if the diameter is known to us.

In order to obtain the true hardness number, the thickness of the specimen should be at least 7 times the depth of the impression, so that for thin work a 1mm ball should be used instead of the more usual 10 mm or 5 mm. sizes.

Various Uses of the Brinell Test. Following under given are some uses of Brinell tests:

(a) To test the hardness, enabling the effect and uniformity of heattreatment to be verified. The amount of hardness induced by cold working may also be examined.

(b) To explore machinability. Depending on the material of which the cutting tool is made, the Brinell number gives a guide as to the machining possibilities. With ordinary cutting tools machining becomes difficult when the hardness number is above 280-320. Below 120-100 a material may be too soft and may tear under the cutting edge, particularly for such operations as screwcutting, milling and broaching.

(c) To estimate the tensile strength (of steel). There is a definite relationship between Brinell hardness and the tensile strength of steel. For annealed (normal) steel:

tensile strength (N/mm^2)=3.2x Brinell number

For hardened and tempered steel:

tensile strength (N/mm^2)= 3.2×Brinell number

(d) The form of the ball impression given an indication of the workhardening capacity of the material.

Piling up (Fig. 4) denotes a low capacity for work-hardening, while *sinking* indicates a high work-hardening capacity.

Table 2 shows the Brinell hardness number for various materials.

Hard Materials Testing

If Brinell number is exceeding about 550, than the Brinell test is not appropriate because the material becomes comparable in hardness to the ball and the latter tends to platten. While we test such type of material, other type of method is being used.

Table 2.2: Brinell hardness values

Material	*Brinell number*
Soft brass	60
Mild steel	130
Annealed chisel steel	235
White cast iron	415
Nitrided surface	750

The Vickers Pyramid Test

When we want to test the high degree of hardness than Vickers pyramid test is provide balanced and appropriate results. It employs a pyramid-shaped diamond having an angle of 136° which is pressed onto the surface with loads ranging from 5 to 120 kgf, the particular load used depending on the hardness range of the material being tested. The diamond makes a small square impression and the Vickers pyramid hardness number is found from tables after the reading across the corners of this impression has been obtained by a microscope fixed to the machine. This method is very appropriate for testing finished work, as the impression made is very minute and does not mar the surface of the component.

The Rockwell Test

In the Rockwell test we use the diamond with 120° angle loaded with 150kgf (scale C) or asteel ball 1.6 mm diameter loaded with 100 kgf (scale B). The ball or diamond is first loaded with 10 kgf to take up the backlash (slack) in the machine, and the indicator is set to zero. The main load is then applied and, after its removal, the measuring device on the machine records the depth of penetration in terms of the Rockwell number.

The Shore scleroscope included essentially of a graduated glass tube inside which a small diamond-pointed hammer may move freely. The hammer, having a mass of 2.37 grams, is allowed to fall from a height of 254 mm onto the surface to be tested, and the height of the first rebound is taken as the index of the hardness. This method is useful for large articles which cannot conveniently be placed under one of the other testing machines since, if necessary, the scleroscope may be carried to, and rested on the article to be tested.

Hardenability (Jominy) Test—BS 4437

This test can be defined by the extent of hardening that occurs when a steel is quenched from the austenitic condition. The test consists of heating a cylindrical test piece to a specified temperature in the austenitic range and water quenching one end. After suitable preparation of the surface, the hardnes along the test piece is determined at intervals from the quenched end.

The Jominy test is shown in the Fig. 5 (a) and alternatives is also given in BS 4437. The test is conducted by heating the test piece quickly and uniformly to the temperature specified for the steel under test, and maintaining it at this temperature for 30 minutes. Precautions must be taken to avoid carburisation or decarburisation and to keep oxidation and scaling to a minimum. When the heating is completed, the test piece is transferred quickly (5 seconds maximum) to a fixture which provides vertical support for it and which incorporates a water connection and jet whereby water is directed onto the lower end of the specimen for quenching – see Fig. 5(b). The jet of water rises to a free height of 65 mm above the orifice without the specimen in position, and the quenching action is allowed to continue for 15 minutes. After quenching, flats to the depth of 0.4 mm are ground on opposite sides of the shank, and hardness readings are taken on these flats at intervals along the length. These may be Rockwell C or Vickers determinations, and the positions at which readings must be made are as follows (measured from the quenched end of the test piece):

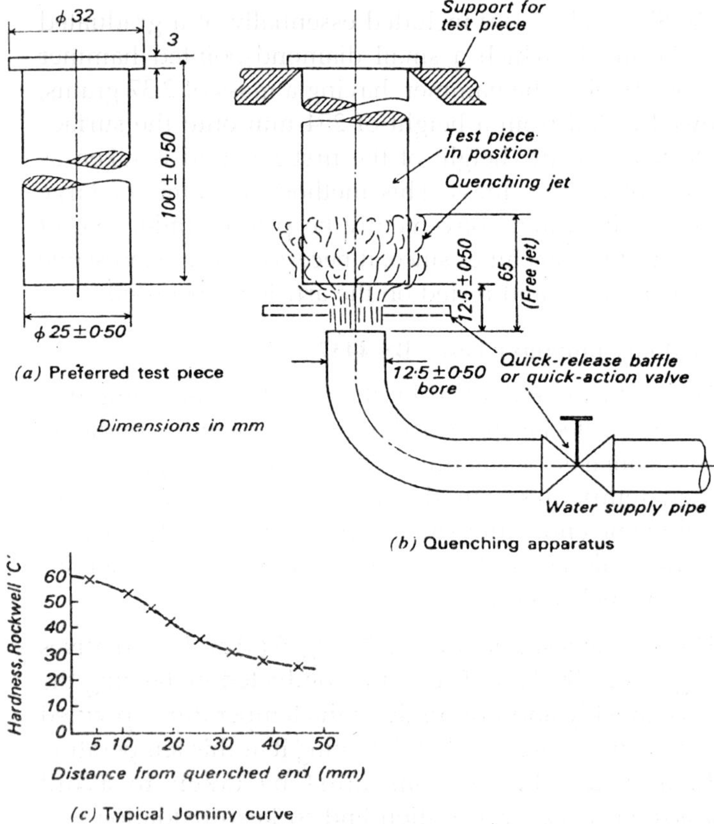

Fig. 5 : The Jominy Test

1.5 mm, 3 mm, then at 2 mm intervals to 15 mm, followed by 5 mm intervals thereafter to about 50 mm.

The mean hardness values for each of the two flats are then plotted vertically to a horizontal scale of distances. A typical graph is shown is Fig. 5(c).

The graph helps in determining the index of hardenability, which is designated by a code indicating the distance *d* from the quenched end of the test piece within which the specified hardness is to be obtained. Thus J38/9 means that the requirement is for a hardness of 38 Rockwell C at a distance of not less than 9 mm from the quenched end.

THE IMPACT TEST

This test is carried out on a specific specimen which is in the form of a notched bar. This testing machine contain essentially of a pendulum which suspended above and swings across a vice or lest which carries the specimen. The pivot of the pendulum is carried on frictionless bearings and in the block which forms the body of the pendulum is incorporated a hardened steel anvil, or striker, which hits and fractures the specimen. In the Fig. 6 (a) we are showing a machine, in this machine the pendulum, indicating dial, operating levels and specimen is indicated.

The LH lever releases the pendulum which the other is for clamping the vice. The *Izod* test uses one or other of the specimens shown in Fig. 6 (d). The specimen is clamped in the vice and set by a gauge with the end notch level with the face of the vice jaw and a length of 28 mm projecting [Fig. 6(b)]. The pendulum is then raised to a predetermined operating position in which it stores up 160 joules of potential energy by virtue of its weight and height above the point where it strikers the specimen. It is then released, swings down, strikes and fractures the specimen and swings past some distance on the other side. The height to which it swings over depends upon how much energy has been absorbed in fracturing the specimen; if no energy were taken it would swing to its original height, while if all its energy were taken it would swing to its original height, while if all its energy were taken it would be arrested at the bottom of its swing.

Through the pointer on the indicating dial, the amount of energy absorbed is shown here. The notches in the specimen are milled with a special form-cutter and specimens are usually made long enough to allow three tests in different directions. The *Izod* value is expressed in joules (Nm) and represents the energy necessary to fracture the test piece.

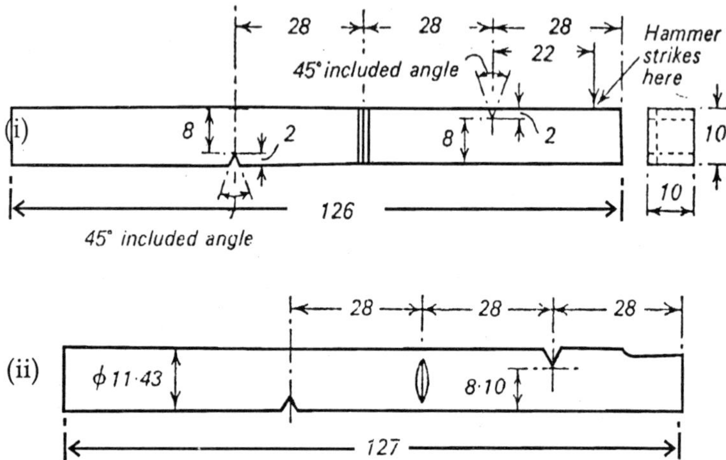

Fig. 2.6: Notched-bar impact test. *Izod* specimens: (i) square and (ii) round (dimensions in mm).

In the *Charpy* test, a 10 mm square specimen with a central notch is held on a rest between supports 40 mm apart and centered [Fig. 6(c)]. The striker on the pendulum is in the form of a V, the sharp edge of which strikes the centre of the specimen after the pendulum has fallen from a height at which its stored energy is 294 joules. The Charpy test is more convenient when the specimens are in the heated or frozen condition, since the setting of the specimen is more quickly performed, and there is less chance of any significant change in temperature taking place. In addition to the above transverse tests, impact values may be obtained on specimens in tension. The specimen has a test diameter of 6.35 mm and is screwed M 12 × 1.25 at the ends, which are carried in special adaptors.

The main purposes of the transverse impact test are as given below:

1. It gives a guide to the resistance of the material against stress concentration at a change in section;

2. It indicates the resistance of a material to the spread of a crack after it has once formed. A low Izod value indicates that in service there will be a greater chance of failure before the initial crack is discovered.

The use of this test to indicate the resistance of a material against impact is not now considered to be reliable, since the speed at which facture takes place is too low. Heat-treatment and tempering are very influential on the Izod value for steel, as Table 3 shows.

TABLE 2.3: Effect of various tempers on the Izod value for a carbon steel. C 0.43–0.50%, Mn 0.60-0.90% (steel previously hardened)

Tempering temperature (°C)	Izod value (joules)	Brinell hardness
320	20	311
430	31	277
480	41	262
540	52	241
600	65	223
650	81	207
700	95	196

The effect of grain direction introduced by rolling causes a higher Izod value to be given when the test is made across the grain than when the specimen is fractured parallel to the grain direction.

TESTS ON SHEET METAL

In the industrial units we have basically used sheets and strips extensively, components, and particularly press-tool work are made by them. In the process of above given functions the material is often subjected to binding and drawing operations. To examine the suitability of material for such type of purposes the following under given are certain tests. Students meets carefully read them.

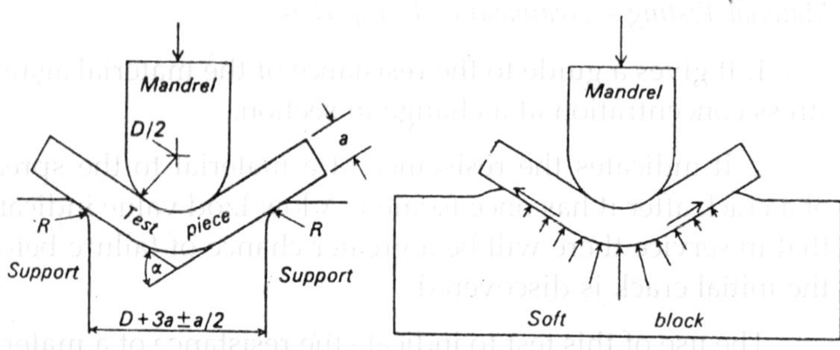

(a) Simple method of bend test (b) Bending on a block of soft material

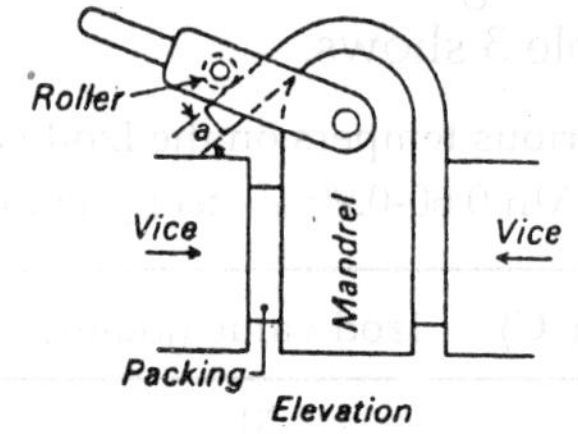

(c) Bending in a bend test machine

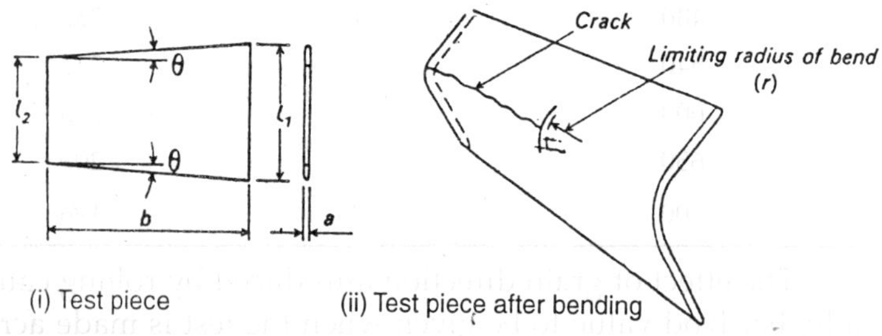

(i) Test piece (ii) Test piece after bending

(d) Cone bend test

Fig. 2.7: Single-bend test

Bend Tests (BS 1639: 1964)

On the basis of BS the specification can be classified in two types: (1) Single-bend tests and another is (2) Reverse-bend tests. In the single-bend tests the piece is submitted to appreciable deformation by a single bend. On another hand reverse-bend tests can be used, where the test piece, held firmly in a vice is alternatively bent and straightened over a mandrel of specified radius. For the purpose of the tests the thickness of materials is graded as (i) *thin* material up to 3 mm thick, (ii) *medium,* material 2.5-12.5 mm thick, and (iii) *thick,* material over 9 mm thick.

Single-bend tests

Method 1: Mandrel and supports. Under this method the test piece is laid on two parallel supports and bent by pressure applied by means of hardened mandrel or former, the radius of which is specified in the standard for the material [Fig. 7(a)]. To accommodate different thicknesses of test piece, the opening between the supports should be $D+3a\pm a/2$. The single-bend tests method is appropriate for testing medium and thick materials for angles of bend up to 120°.

Method 2: A variation of Method 1 employs a vee block as the support and is suitable for bends up to 90° in medium material.

Method 3: Support of soft material [Fig.(b)]. Here the test piece is pressed into a block of lead or other suitable plastic matrix by means of a hardened mandrel. This method of material testing can appropriately used for testing this materials for bends up to 90°. For greater bends than 90° the legs of the test piece may be closed by side pressure suitably applied.

Method 4: Bending round a mandrel [Fig. 7(c)]. the test piece is held against a mandrel in a vice or clamping device and is bent around the mandrel by applying a force at right angles to its surface. Bending may be effected by a machine as in Fig. 7(c) or by blows with a mallet. If a machine is used, the force should be applied at a distance not greater than $2a$ from the last line of contact between the test piece and the mandrel. In case of this and medium material for bend up to 180° this method can be used appropriately.

Method 5: Free bend. The test piece, which has been initially bent by Method 1 or 2, is subjected to compression between jaws until the angle of bend attains the specified value. Free bend test generally do not control the final radius of the bend. That is why it is suitable for medium and thick materials for bends up to 180°.

Method 6: Pressure bend. The procedure is similar to the free-bend test, but control is exercised by bending over a former or continuing the pressure until the inner faces of the test piece come into contact. This method is suitable for all thicknesses of material. Methods 5 and 6 are difficult to control and are should be exercised against accident hazards.

The criteria for the above tests are that, after the piece has been bent, it shall be unbroken and free of cracks on the outside surface of the bend. Hair cracks at the edges of the test piece may be disregarded.

Method 7: Free-cone bend. Under this method the test piece is shaped to a trapezium and then bent under pressure between a pair of shaped jaws. Fig. 7 (d) describe the analytical Fig. of cane bend method. This loading produces buckling and bending to a progressively decreasing radius from the broad to the narrow end. The test-piece dimensions are chosen so that the radius at the narrow end is less than the limiting radius of bend, and cracking occurs as shown at (ii) in Fig. 7 (d). Measurement of this limiting radius at the end of the crack provides the limiting value for the material. This mettod is suitable for the 'thin' and 'medium' material up to 90^0 bends.

Reverse-bend Test

*Method 8 (Fig. 8).*Under this method one end of a rectangular test piece is held between hardened jaws with their inner edges formed to a specified radius. The free end of the test piece is bent is bent through 90° in opposite directions. One reverse bend consists of taking the end through 90° and then back again to its original position. In this test, the material must be provided to ensure continual contact between the test piece and the reduced edges of the jaws. By applying tension or by rollers this may be achieved. This test is used for 'thin' materials until a specified number of bends has been achieved or until the test piece fails.

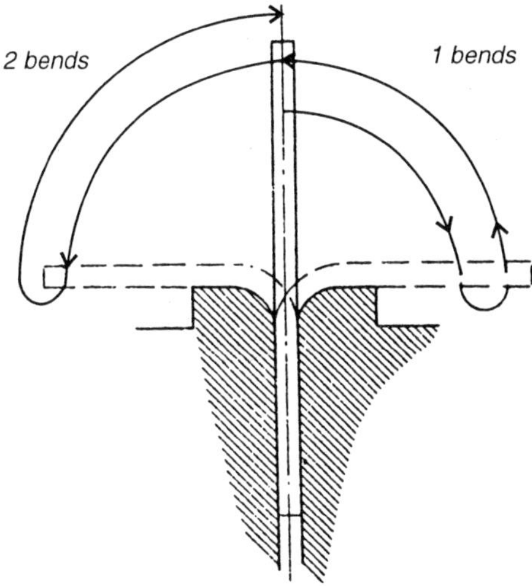

Fig. 2.8: Reverse-bend test

The Erichsen Test

This method can be applied on the sheet and strip of both ferrous and non-ferrous nature having thicknesses from 0.1 mm to 2 mm inclusive. The material is gripped between two faces under a load of 1000±100 kgf and a 20 mm ball or spherical penetrator is pressed onto it, causing its surface to be drawn down into a dome as shown in Fig. 9(a). At a certain point, depending on the

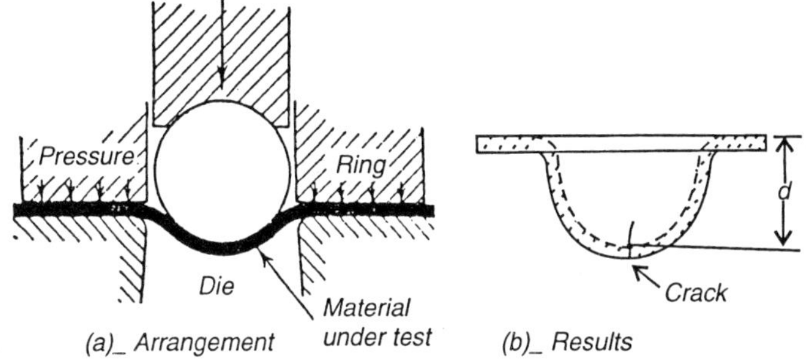

Fig. 2.9: The Erichsen test.

material, the top of the dome will crack (b), and at this point the test is completed. The depth *d* mm to which the dome has been pressed when the crack appears is the Erichsen number (IE), and this gives an approximate measure of the relative capacity of different materials for being worked and drawn into cup-shaped articles. The surface appearance of the dome also gives an indication of the original grain size of the material and the probable appearance of the surface after it has been through the press tools.

Elasticity and Plasticity of the Metals

When we provide force to a metal it tend to be strained or distorted. The amount of distortion or strained is depending on the severity of the force producing it. Generally metals in their cold state behave in an elastic manner and if the force provided is not greater than the elastic capacity of the materials to withstand it, than the metal returns to its original shape when the force provided is taken away. Such actions leave the structure of the metal unaltered. When, however, the force is great enough to distort the metal beyond its elastic capacity a permanent deformation is produced which does not disappear on the removal of the force. A permanent deformation of this nature involves some plastic flow in the material and causes changes in its internal structure which will be discussed later.

In contrast to the designer, who deals with materials loaded within their elastic range, the workshop engineer is much concerned with the plastic flow of metals, since in all cutting, pressing, forging operations, etc. the material, or some portion of it, is put into this condition.

METALS—WORK-HARDENING

When a material is deformed permanently, within it plastic flow takes place and other many changes occurs in the structure of it, basically such as to make metal harder. The work performed in the workshop are generally of this nature and such operation is carried out when the metal is cold. i.e. below the lower critical point came under the category of cold work. These work included cutting, binding and forging, etc.

Severe cold working renders most materials harder and less ductile than they would be in their normal annealed condition. If a large amount of cold working has to be performed, the hardening effect may be sufficient to prevent more than a certain amount of change in shape taking place, and the material has to be annealed before the process may be continued. This occurs in the drawing of wire, the cold rolling of strip, etc. where, at certain stages in the reduction, the material must be annealed to bring it to a suitable condition for further working. The effect of directional working (e.g. rolling) is to elongate the grain structure of the material in the direction of the distortion, and Fig. 10 shows at (a) the normal structure of a mild steel and at (b) its structure after severe cold rolling. It also shows the microstructure of the material at the edge of a sheared plate (c).

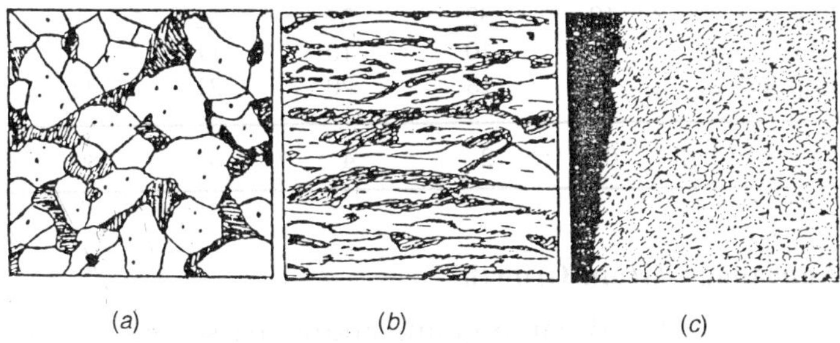

(a) (b) (c)

Fig. 2.10: Effect of cold working on the structure of a metal. (a) Normal structure of mild steel (0-17%C). (b) Same steel after a considerable reduction by wire drawing. (c) Structure at edge of sheared plate (×100).

To enhance the hardness and tensile strength, to reduce the ductility and to make the metal more brittle are the general effects of the cold working. These effects may be completely removed by annealing, but for steel most of the effects may be removed by heating to 600°C, soaking at that temperature and then cooling off in air. This treatment causes recrystallization of the ferrite and removes most of the qualities which might be undesirable in the structure. Fig. 11 shows the relative hardness at various locations on the chip being removed from work by a cutting tool, and demonstrates the severe cold working induced by a cutting tool.

When a material is subjected to increasing amounts of cold work, its induced hardness rises to a certain maximum amount beyond which, if it is worked further, the hardness does not increase and may decrease. The maximum induced hardness which provided by the cold work to a metal is depend on two facts one is the nature of material and another is influenced by the temperature. Table 4 is showing the results of experiments on the work-hardening of various materials, the induced hardness being effected on the surface by rolling with a hardened ball under pressure.

Table 2.4: Induced hardness off various materials

Material	Hardness number (Herbert pendulum)					
Mild steel	21	30	30	31	32	30
Stainless steel	20	41	45	47	48	50
Brass	19	32	32	32		
Aluminium	6	10	10	11		
Copper	7	18	20	21	22	24
Passes of ball	09	2	4	6	8	10

Annealing—Grain Growth

As steel with large grains and a coarse structure tends to be brittle and to give an inferior surface after certain pressing operations, any treatment which promotes such conditions should be carefully controlled with a view to suppressing them. We have seen that prolonged heating above the upper critical point tends to coarsen the structure of steel,* and at this point we might mention another cause of this fault. When steel has been cold worked, it is usual, as we have discussed above, to anneal the material in order to rectify the structural distortion. It is possible, by annealing after cold working, to obtain conditions favorable for the formation of large grains, and such an occurrence will take place when the amount of cold work performed is in a particular relationship to the subsequent annealing temperature. This is shown in Fig. 11(a), which shows a piece of steel into which a hardened ball has been forced, after which the specimen has been annealed. The effect of the ball pressure has been to work-harden the material in the neighborhood, intensely at the surface and by lesser amounts towards the interior, until fairly deep down no effect of it will result. The subsequent annealing, operating on material having varying degrees of cold work, has selectively caused grain growth to occur in that portion where the relation between the amount of work and the annealing temperature is critical for the promotion of large grains. In the diagram it will be seen that the intensely worked metal and the unworked metal are left with a fine structure, but at an intermediate layer much grain growth has taken place. The curves at Fig. 11(b). Show the relations between the amount of work, the annealing temperature and the grain size for the material to which they refer.

From this diagram it will be seen that if correct annealing temperatures are used there is little danger of producing a coarse structure, but the effect is possible with incorrect temperatures.

Work-hardening and the Effect of Temperature

Work-hardening and the temperature is corelated. These should be a certain effect on the process with variable temperature. The temperature at which work hardeing is carried out influences the work-hardening capacity. It has been illustrated in the figure 12. The values and horizontal spacing of these six points will depend on the material and, although they are all more or less evident for most materials, they may not all be very clearly defined and some of them may occur at temperatures so low as to be off the left-hand side of the graph. One fact, however, is fairly consistent for all metals: the second depression (D_2) is generally the lowest and occurs at a temperature of approximately 120-140°C. This means that at the D_2 temperature the work-hardening capacity of a material is at a minimum, and if we could do our operations on the metal with it at that temperature we should save in tools, power and spoilage due to cracking, etc.

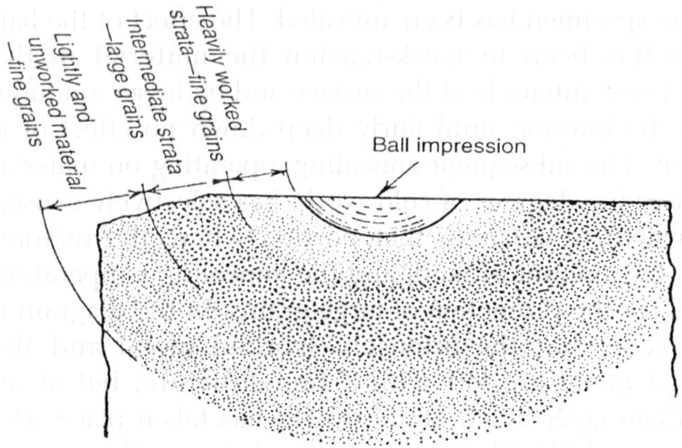

Fig. 2.11: (a) Effect of annealing on grain size after varying amount of cold work.

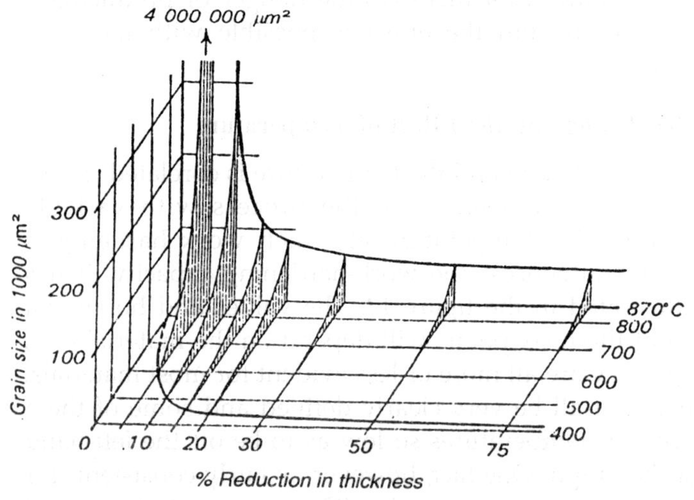

Fig. 2.11 (b) Relationship between cold work, annealing temperature, and grain size for iron.

The Whitaker ring is an interesting confirmation of the existence of the D_2 temperature effect. When a bar of material has its end faced up in the lathe at a constant speed, the cutting speed increases from zero at the centre to a maximum at the outside diameter. The temperature at the tool point will likewise increase from the centre to the periphery. At some intermediate diameter a cutting temperature is obtained which corresponds to that at D_2 for the material being machined, and under favourable conditions a bright ring is produced on the faced surface.

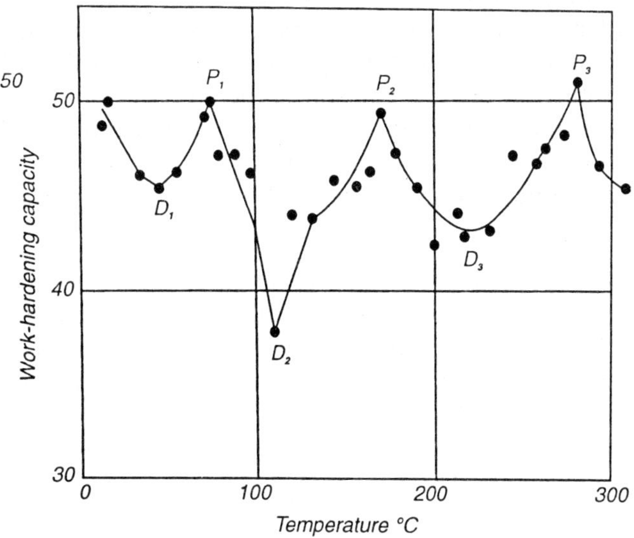

Fig. 2.12 Temperature-work-hardening curve of a steel.

HIGH-SPEED STEEL (HSS)

High speed steel is widely used the manufacturing sector. It is also known as 18-4-1 variety, this unique name is because of its unique composition. It contains 18% tungsten) 4% chromium, 1% vanadium, 0.6-0.8% carbon, and remainder iron. During the 1939-45 war, the loss of tungsten supplies compelled the use of alternatives, and it was found that in the constitution of high-speed steel molybdenum could be substituted for some of the tungsten without much loss of efficiency. This resulted in the development of the so-called 'substitute' steels, the two chief ones being substitute '66' and substitute '94', the numerical designations meaning that the

66 quality contained 6% each of molybdenum and tungsten, while the 94 quality contained 9% molybdenum and 4% tungsten. Before their introduction into Great Britain, these steels had actually been developed in America, and it is claimed that with proper manipulation during heat-treatment they are comparable with the previous qualities of HSS. Their greatest drawback was the proneness to decarburization and scaling during heat-treatment, and at the time this resulted in the wider use of heat-treatment processes and precautions which excluded oxygen during heat-treatment. Although, with the return of tungsten supplies, the employment of the substitutes has declined, the heat-treatment methods developed for their use still remain and give improved results even with normal tungsten steels. We shall discuss these directly.

The growth of different heatresisting, corrosion resisting and high strength alloys to satisfy the conditions found in nuclear, aircraft and gas-turbine industries has make machining problems beyond the capacity of previous qualities of highspeed steel. This has led to further work on cutting alloys and an increase in the amount of vanadium with cobalt as alloying elements. The alloys so produced have improved properties of hot-hardness, wear-resistance, etc. at the expense of some toughness, to cope with the arduous machining conditions involved. The 10% cobalt alloy has cobalt 10%, tungsten 20%, chromium 5%, carbon 0.5%, and is suitable for heavy, continuous turning. The high-vanadium steel has the composition 4% vanadium tungsten, 4.5% chromium and 1.2% carbon; while the vanadium-cobalt alloy has 5% each of vanadium and cobalt, 12.5% tungsten, 4.5% chromium and 1.5% carbon. These steels have high wear and abrasion-resistance and are suitable for tools and milling cutters under the most severe cutting conditions.

As high *HSS Furnaces* speed steels should essentially be heated to a temperature of the order of 1300°C, and at such a temperature are prone to scaling and precautions should be taken while excluding oxygen from the process. Methods in current use employ either a salt-bath furnace, where the steel is heated by immersion in a molten salt, or a 'dry' muffle-type furnace. Salt-bath furnaces prevent the surface defects we have mentioned, since the article is immersed in the fluid of the bath and so protected from the agents (air, gas, etc.) which cause scaling. When a 'dry' muffle-type furnace is used, it is necessary to make provision for the maintenance of an atmosphere inside the furnace which mitigates against decarburisationn and scaling and prevents the ingress of air which would cause this to take place. Such furnaces are called *controlled-atmosphere* furnaces. The hardening of highspeed steel is having a preheating stage in which the work is carefully heated to about 850°C. After that the work will be transferred to the another furnace, in which it must be rapidly brought to the hardening temperature. For normal high-speed steel this is about 1250-1300°C, while for the substitute steels a temperature range of 1220-1280°C is suitable. According to the maker's instructions the steel, from its hardening temperature, should be cooled off in air, or quenched in an air blast or oil bath.

The high-speed hardening furnace. Electric high speed hardening furnace is a muffle type for treating highspeed steel are generally of a doublechamber type. One chamber is operating up to 850-900^{0}C in which the preheating takes place. And on the other, chamber provide heat enough to bring the steel to the hardening temperature. Heating is usually carried out by gas or electricity, the high temperature in the gas furnace being obtained by feeding the air under pressure from a blower. In these furnaces the low-temperature unit is often heated by the waste gases from the high-temperature chamber. Pyrometric arrangements are made for temperature indication and for its automatic control if required. A suitable atmosphere inside the furnace is maintained by special methods for the control of the mixture of gas and air for combustion, by feeding an auxiliary supply of suitable gas, or by both. Electric furnaces must, of course, have means for an independent supply of protective gas.

Salt-bath furnaces. In this process the component is heated by being immersed in a bath contain liquid salt i.e. sodium cyanide. The solid salt is being used for these furnaces at ordinary temperature, but often temperature (heated) is sufficient for heat-treatment than they melt and become viscous liquids. Salt-bath furnaces may be of the high-temperature type suitable for heat-treating high-speed steel, or a more common type, operating at temperatures up to 1000°C, may be adapted for heating components preparatory to quenching, and, when used with a suitable salt, for carburising a 'case' of high-carbon steel as an alternative method to that in which the components are packed and heated in contact with a carburising compound.

In the high-temperature salt bath used for high-speed steel, the salt is heated electrically by passing current between two electrodes of pure iron immersed in the salt and fed at low voltage from a transformer. The molten salt, while conducting current between the electrodes, possesses sufficient resistance to promote the heating necessary to operate the furnace. Special arrangements are necessary for starting the furnace until the salt melts and is able to conduct current between the main electrodes. A carbon rod fixed between two auxiliary carbon electrodes, is supplied with current and glows red. This melts the salt, which increases in conductivity as it becomes liquid, and further salt is added until there is enough molten salt for the main electrodes to be used. When the furnace is under way, the auxiliary electrodes are removed and at the end of the day's work the salt is baled out for use again. A diagram of an electrode salt-bath furnace is shown in Fig. 3.

Salt-bath furnaces are made with bath dimensions from 100 mm × 110 mm × 150 mm deep to 250 mm dia.×300 mm deep, taking discs from 90 mm dia.×13mm thick to 200 mm dia. ×25 mm thick or long work from 38 mm dia.×110 long to 125 mm dia.×230 mm long, with nominal ratings of 5 kW to 12 kW.

Furnaces of this type may be operated over a temperature range of approximately 600°C to 1400°C.

Secondary Hardening—Tempering

After quenching, the steel will surely tempered here the highspeed steel demonstrate most interesting results. As you well known that when carbon steel is tempered after hardening, it progressively loses its hardness until at about 750 °C it has softened to its normal state. With high-speed steel, temperatures up to about 250°C have practically no effect on the microstructure or hardness. From this temperature to about 450°C the hardness drops slightly and toughness increases. From 450°C to 600°C the hardness rises again and may reach a value higher than that reached by the quenching. This is known as *secondary hardening,* and it is thought that the increased hardness comes as a result of retained austenite from quenching gradually breaking down into the harder constituent, martensite. (The secondary hardening temperature, for substitute steel is about 550°C.)

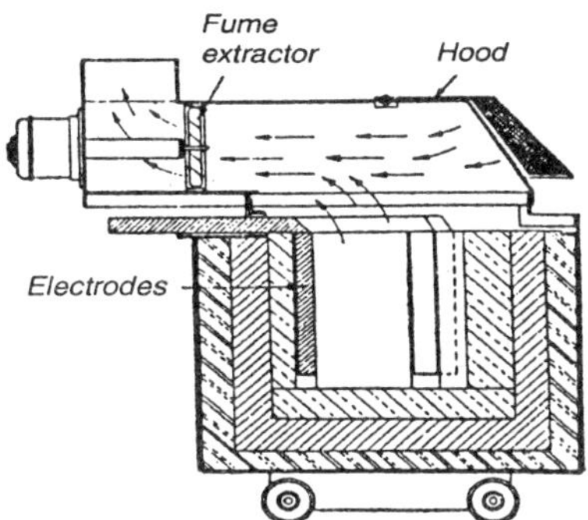

Fig. 2.13: Section.

The crucial factors in the heat-treatment of steel are temperature and time. If it is held at its hardening temperature appreciably longer than the time necessary for uniform heating, grain growth occurs rapidly and may render the steel coarse and useless (hence the need for rapid heating for 850°C to the hardening temperature). Unless it is hardened from the correct temperature it will not have the best structure, the secondary hardening quality will not be brought out and the 'red-hardness' property will not be present.

Normal Salt Baths

Normal-type furnaces have a steel pot consists of the salt which is heated externally. These furnaces are made in various sizes and shapes and are used for a wide range of heat-treatment processes on ferrous and non-ferrous metals.

Case-hardening

To provide a hard, wear-resisting 'case' combined with a tough, ductile core is a crucial one and is used for various moving parts in mechanical assemblies. In the previousbook we have already discussed the method of carburising work by packing it in a box with a carbon-rich compound, heating until the surface skin has acquired a higher carbon content and then heat-treating this for hardness. For many years this method served us well enough, but it has several disadvantages and by modern standards it has become a rough and ready method of achieving the object. It is, therefore, rapidly being superseded by other methods which lend themselves to more refinements in control and quality, can be fitted better into production flow schemes, obviate the dirty and laborious preparation and clearance of the work and compound from boxes, and give economies in fuel, labour, materials and floor space. In addition, the aesthetics of the newer processes are more in keeping with modern standards.

Gas carburising

We known that all the equipment being installed today for carburisation uses the principle of the carbon enrichment of the component surface by immersing it in a carbon-rich atmosphere. Carburising atmospheres for this purpose usually contain the constituents H_2, H_2O, CO, CO_2, CH_4 and N_2, and there are two principal methods for producing the atmosphere. In the first method of these a neutral or mildly carburising carrier gas generated from town gas or liquid petroleum gas is enriched with a small percentage of hydrocarbon and fed into the furnace containing the heated components. A typical composition of the carrier gas is CO

20-25%, H_2 40-45%, CO_2 0-1%, CH_4 1-2%, balance N_2. The amount of added hydrocarbon (butane) may be up to 5%.

The second method of obtaining the furnace atmosphere is known as 'liquid-fed' or 'drip-feed' method. In this a suitable organic liquid, such as a mixture of isopropyl or methyl alcohol with benzene, is dripped directly into the furnace, where it cracks to form an atmosphere of typical composition CO 27%, H_2 49%, CH_4 21%, CO_2 0.4%, N_2 balance. Atmospheres of this type are less controllable than the carrier-gas type but are suitable for many operations.

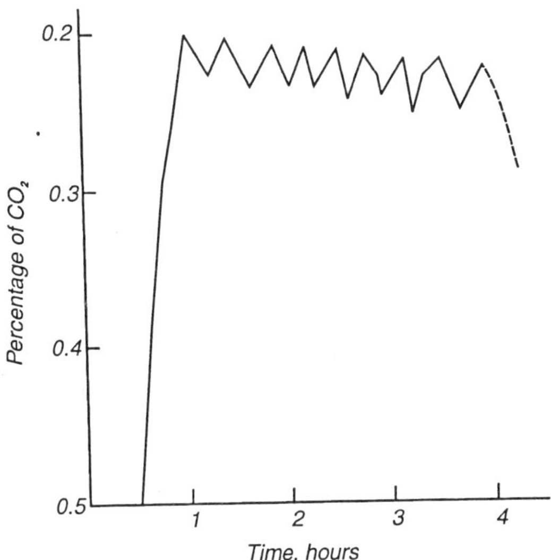

Fig. 2.14: Kent Grubb-Parsons infra-red CO_2 control of carburising cycle.

On the careful control of the carburising atmosphere, the efficiency of carburisation is depend. Carbon potential is inversely proportional to carbon dioxide and to water content, so it may be controlled by regulating either the CO_2 content of the dew-point. Formerly, the control was on the dew-point, but this is being superseded by a system using infra-red analysis of the CO_2 content of the gas. The basic equipment includes an infra-red gas analyser for CO_2 content of the gas. The basic equipment includes an infra-red gas analyser for CO_2 values, a recording instrument and a controller. The range of the recorder and analyser is 0 to 1% CO_2, and the speed of scanning about 1 minute. A portion of a recorder reading is shown in Fig. 14.

Case-hardening requirements are include the following: (*a*) a specified hardenable depth of case, usually with an allovable variation of 10-20%, and (b) a surface carbon content within the range of 0.7-1.1%, with a permitted variation up to 10%. In the furnace the control may be exercised in two ways: (*a*) maintaining the carbuising atmosphere throughout the cycle, or (*b*) maintaining an 'active' carburising atmosphere for a portion of the cycle and then following with a period of 'diffusion' in an atmosphere of reduced carbon potential when the high carbon content at the surface diffuses inwards. The case depth depends entirely on time and temperature, the relationships in question being shown Fig. 15.

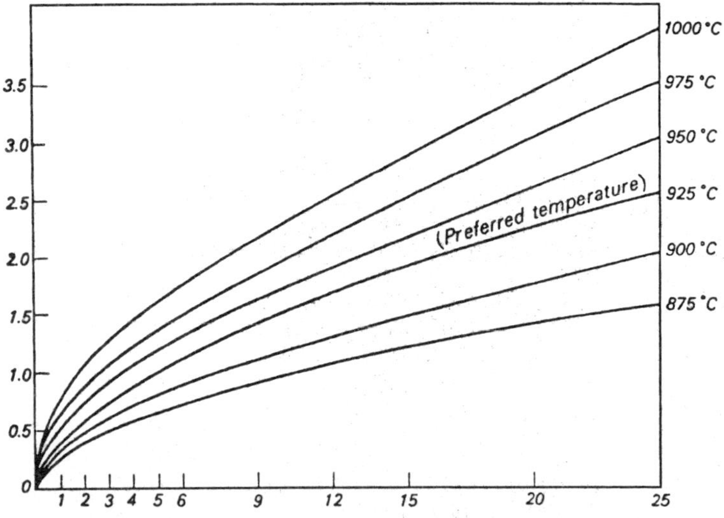

Fig. 2.15: Time-penetration curves for gas carburising.

Required Equipment. The following basic elements are require for gas carburising: (*a*) a heated work chamber effectively gas tight when closed, (*b*) equipment for maintaining the carburising atmosphere in the work chamber, (*c*) means to expose the furnace charge uniformly to the atmosphere gas, (*d*) a furnace-firing system which neither affects nor is affected by the atmosphere inside.

In addition it will be necessary to provide means for quenching or for cooling the charge while protected from the atmosphere.

Furnaces are of three general types:

1. pit type into which the charge is lowered in fixtures or baskets (Fig. 16);
2. horizontal type into which the work on trays or baskets is pushed on rails or rollers;
3. rotating retort type which can be charged from a hopper and tilted to discharge the load.

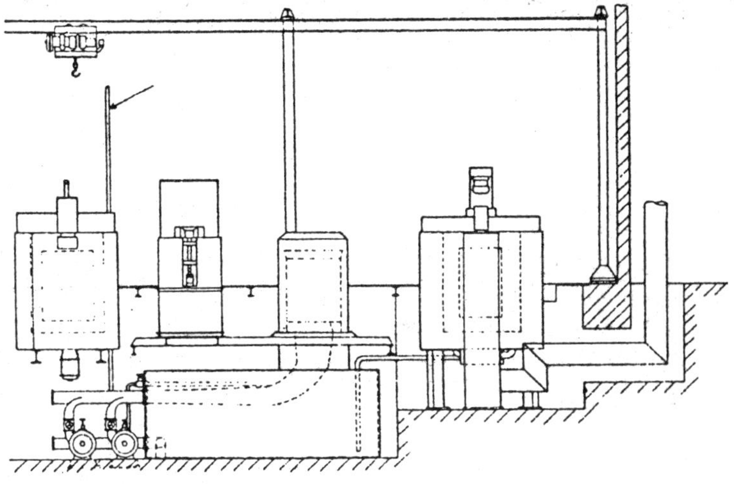

Fig. 16: After-treatment equipment associated with a pit furnace

There are in addition, different specially designed furnaces and equipment for continuous operation or functioning.

For long components, the pit type is suitable such as camshafts, and the retort type for small components being handled in bulk. For the first two types a fan is necessary to ensure efficient gas circulation and unifourmity of carborisations but this is achieved in the retort furnace by the tumbling action of the furnace contents.

The working chamber of furnaces can be heated by electricity or gas but the requirement are necessary, which are (*a*) above necessitates special arrangements. In the smaller furnaces the muffle is lined with a heat-resisting alloy which is externally heated by electric elements or gas burners. Larger furnaces have a gas-tight outer steel casing, and this is heated by enclosed gas-fired or electric radiant tube heaters around its periphery The design of the horizontal furnace lends itself well to the incorporation of integral quenching arrangements, so that the work may be lowered into the quench without exposure. In the furnace shown, the square opening is the door of a sealed vestibule beyond which is the heating chamber. The upper part of the vestibule houses the elevator gear for lowering the charge container into the quench tank housed in its base. Pit furnaces are made in sizes roughly ranging from 500 mm diameter by 1 metre deep (40 kW) to 1½ metres diameter by 3 metres deep (gas heated), while the range of horizontal furnaces goes from 70 litres to 600 litres capacity. A diagram showing the after-treatment equipment associated with a pit furnace is shown in Fig. 16.

Nitrogen hardening

It is a specific method of case-hardening by a gas process. In this process the ammonia (NH_3) is used with a special steel already in the hardened and tempered condition. It results in the absorption of nitrogen and the formation of nitrides of molybdenum, aluminium and chromium in the case. The result is a thin and intensely hard case upon which no further treatment, e.g. quenching, is necessary. The carbon content of nitriding steels varies from 0.2 to 0.5% and in addition there is aluminium (up to 1%), chromium (about 1.5%) and molybdenum (0.2%). The parts to be treated are loaded into a heat-resisting steel box with inlets and outlets for the gas and loose packing (e.g. nickel gauze) to separate the components. The box is then charged into the furnace, brought up to the necessary temperature and maintained there while a steady stream of ammonia (NH_3) is passed through. The working temperature is 500°C and an essential requirement is a close adherence to this (±5°C). The case builds up approximately at the rate of 0.2 mm per 20 h of heating up to a thickness of about 0.75 mm, after which it is slower. After the treatment no further hardening is necessary. Parts for nitriding must be chemically clean and bright, and any stressed condition (e.g. work-hardening by maclining) is removed by the preliminary heats-treatment. If any part of the surface is required in the soft condition, nitriding can be prevented by plating with copper or tin. During the process a slight growth occurs and an allowance of about 0.025 mm must be made.

Following are certain advantages of nitrogen process:

(*a*) it employs a temperature below the critical range, so that distortion is minimized or non-existent, and in some cases subsequent grinding is unnecessary;

(*b*) the absence of quenching removes the risk of cracking;

(*c*) the case is extremely hard and wear-resistant, and retains these properties up to about 500°C.

Against these advantages must be set the rather long heating time for any but very thin cases and the higher costs of the special steel necessary.

Carbonitriding process

In this process, both carbon and nitrogen are simultaneously absorbed into the case, and in this it resembles salt-bath cyaniding (see next section). The furnaces used are similar to those employed for gas carburising, and any atmosphere suitable for that process can be used with the addition of a proportion of anhydrous ammonia. The respective absorptions of carbon and nitrogen will depend on the concentration of carbon and ammonia in the atmosphere and to some extent on the working temperature. Carbonitriding is usually carried out at 800-900°C and the penetration rate is about 50% above that for straight carburising at the same temperature. This is an important advantage of the process, since it can compete on equal terms with carburising when operating at a lower temperature with its reduced tendency for distortion and lighter demand on furnaces. Another advantage is the increased hardenability of the case, which allows full hardening with an oil quench even for heavy sections. The process can be used for thick cases, but its advantages lie more particularly in the range from 0.1 mm to 0.5 mm thick.

The process of cyanide case-hardening

In this process salt-bath furnaces may be used for carburising the surface of mild-steel components for purposes of case-hardening. For thin cases (up to 0.25 mm) sodium cyanide (20-50%) with soda ash is used, and for deeper penetration (up to 1 mm) barium chloride is added. The articles to be treated are suspended or immersed in the bath and left long enough for the required depth of penetration to take place (1-5 h depending on depth, salt concentration and temperature). After the surface has been carburised, the subsequent treatments to harden the case and refine the core follow the same principles as we have already discussed for box case-hardening.

The process of cyanide case-hardening is clean and quick and the complete immersion of the components enhances the uniformity in heating and penetration and precludes scaling and atmospheric oxidation. When the work is removed from the bath, a film of adhering salt also acts as a protection from atmospheric oxygen but, if quenched straight from the bath, care is necessary against spitting from the adhering molten salt. If local or selective hardening is required, this may be done by immersing only that portion which needs treatment. The method is efficient and well adapted for thin cases (up to 0.25 mm), but is not a convenient one for adapting into any scheme of continuous or flow production. Special precautions are necessary, and there are regulations in force against the poisonous properties of the gases given off from the salt baths.

Induction hardening process

In the previous discussion of case-hardening process we learn that the object is achieved by converting the outer skin of the material to steel of a hardable quality. An alternative method of surface hardening is to heat the surface layers of a hardenable steel to the hardening temperature and then quench it. This results in a hardened outer case and a soft core of the original material. Between the case and the core will be a transition zone with a hardness gradient having qualities depending on the rate and time of heating and quenching. We have already discussed the application of high-frequency heating for melting steel; the same principle can be used for the surface heating of metals as distinct from their melting, and this is achieved by subjecting the metal to the eddy currents induced by a high-frequency alternating current flowing in a coil surrounding the part to be heated. As you known better that steel has a fairly high thermal conductivity, very high heating rates and short heating times are necessary (seconds or fractions of a second) if the effect is to be confined to the surface layers. Induction heating offers the features required, because the eddy currents are generated in the surface layers of the material and, by a appropriate choice of frequency and power input, the required temperature-time relationships can be achieved. The minimum case depths obtainable in normal practice are about 0.15 mm to 0.2 mm when using a frequency of 10 kHz (motor generator supply), and 0.25 to 0.4 mm when using a frequency of 500 kHz (electronic oscillator), the heat input in each case being of the order of 3 kW/cm^2 of heated surface. The most appropriale steel for this treatment is one with a carbon content of about 0.4 to 0.5%, and in practice the process is mechanized so that the component is heated for the required time and then quenched either by water sprays *in situ* or by lowering into a quench tank. The operations are timed automatically and the features of the process facilitate its adaptation to mechanical control.

You know that induction heating is not, of course, confined to surface-hardening techniques, but may be used equally well for heating work right through for forging operations, through hardening, and so on. It is an extremely quick and easy method for dealing with the types of work for which it is suited.

Flame-hardening process

The high-temperature flame is also used to heat work for the purpose of surface hardening but, owing to the lower practicable rates of heat input to the work, the case depth is rather more than when using induction methods. Fig. 17 shows hardness-depth relationships for two carbon steels after flame hardening. Flame-hardening methods are used, in general, for larger components and for deeper cases than would apply to induction heating and, although some measure of mechanisation can be used, it is rather less adaptable than the electrical method. In a flame-hardening machine for the teeth of large gears, the gear is mounted with its axis vertical adjacent to the heating torch and quenching arrangements. In operation, each tooth space in turn is indexed past the flame and then quenched.

Process of gas preparation

In the metallurgy and heat-treatment these is a certain requirement of prepared gas to produce a furnace atmosphere beside these the gas may be used as fuel for heating the furnace. Under given are some examples of such practicees:

(*a*) gas carburisation and carbonitriding, such as we have just discussed;

(*b*) the bright annealing of metals and the coper brazing of steel without a flux, which require neutral or mildly reducing atmospheres;

(*c*) the hardening of high-speed steel, the bright hardening of carburised components, and the sintering of parts made from compressed metal powders. These require a highly reducing atmosphere.

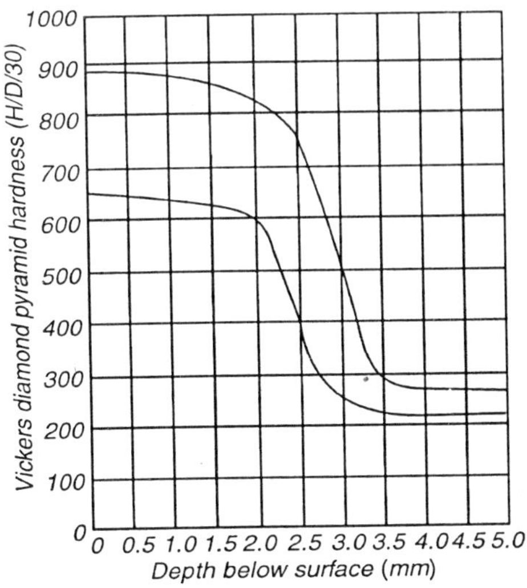

Fig. 2.17: Flame-hardening. Hardness-depth curves for 0.40% and 0.55% carbon steels.

The processes for preparing gases are of two types, given below:

1. *Exothermic* methods, where a gas is burnt with a controlled proportion of oxygen (air) to produce a produce of the required quality. Almost any type of fuel gas may be used, such as town gas, coke-oven gas, bottled hydrocarbon gases (e.g. butane and propane), kerosene and similar liquid-hydrocarbon fuels.

2. *Endothermic* methods, where a gas is passed through a heated retort containing a catalyst which causes the reaction necessary to produce the gas required.

Gas generators are being made with output ratings varying from 7 to 170 m^3/h, and an example of an exothermic generator with its flow diagram is shown in Fig. 18.

Instability after hardening process

For some time after it has been hardened by quenching, carbon steel is in an unstable position and small changes are apt to occur in its shape and dimensions. For articles such as tools, chisels, etc. these small changes may not be detrimental, but, where the finished article is an accurately sized and finished gauge, any changes in it afterwards are likely to be too serious to ignore. As an example, the graphs in Fig. 19 shows, at A_1, the size of a 75 mm length gauge measured at five-monthly intervals after it had been hardened and finished. It will be seen that, when assessed in proportion to modern precision measurement, the chances are considerable and cannot be ignored.

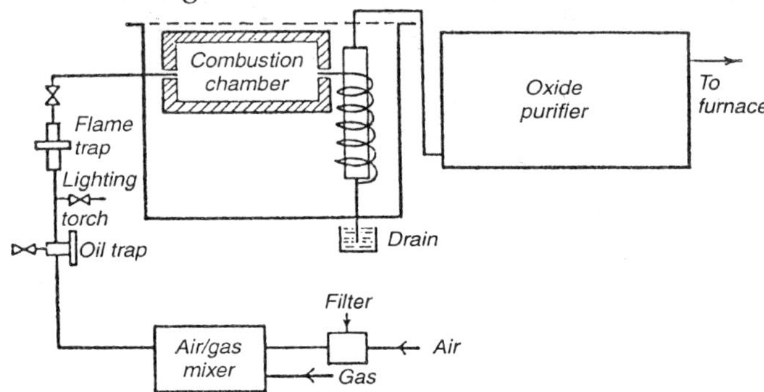

Fig. 2.18: Flow diagram for the Exothermic generator.

To avoid such defect, an ageing period could be allowed to elapse between hardening and finishing off the gauge, but because this might extend to several months such a course in not easy. A more convenient method of ageing is to submit the gauge to 'stabilising' treatment in which it is heated to 150 °C and held at that temperature for about 5 h, afterwards being allowed to cool off. This treatment has been found to be just as effective as several months of natural ageing (Fig. 19, A_2 and A_3).

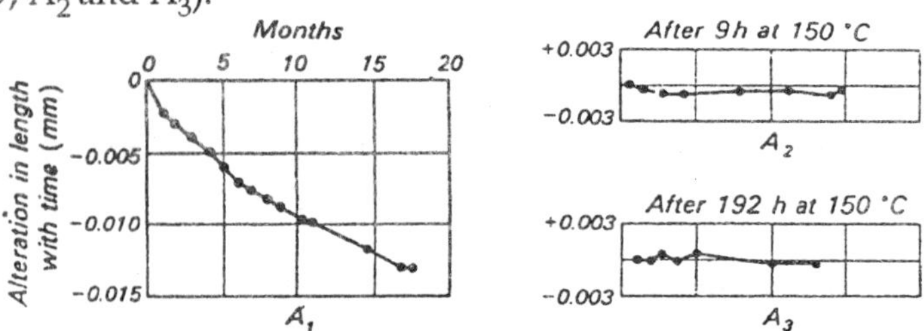

Fig. 2.19: Variation in dimensions of a hardened gauge before and after stabilising.

Hardening cracks

The great misfortune about work that cracks in hardening is that it represents the loss of finished work, often upon which much time and money have been expended. Add to this the dislocation in the planning system when an important tool is lost and must be processed all over again, and we realise that the fault is one which should be minimised as much as possible.

The factors which enhance cracking we have learn previously. It is also a fact that this is a natural phenomena, that is why one can not prevent this; but we can minimize and contract their effects by appropriate manipulation and design.

When steel is quenched, volume changes occur rapidly and unevenly throughout the specimen. The outer skin cools first and may reach its martensitic state before the core has hardly felt the effect of the quench. If the core has still to pass through the critical range, it is due for an *expansion* before its normal cooling contraction, and this expansion must take place inside the already formed hard and contracted skin of martensite. Under such conditions severe stresses set up which may cause the metal either to distort or to crack if the ductility

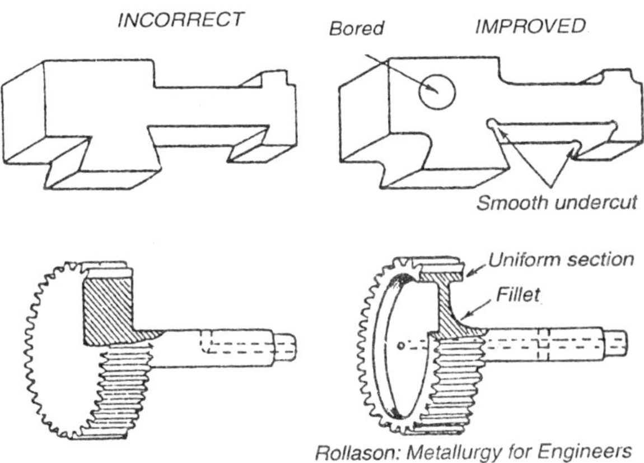

Fig. 2.20: The relationship of design to heat-treatment.

is insufficient for plastic flow to occur. Cracks may occur during or after quenching, or in the beginning of tempering.

We can minimise the distortion and cracking effects by paying attention to the design and by taking precautions during manipulation and hardening of articles which should be water quenched.

The design of such articles should aim at avoiding large isolated masses of metal and maintaining a uniform section as far as possible. Sharp internal corners should be avoided by adding fillets or by undercutting. Particular care is needed when articles are of an irregular section, and Fig. 20 shows two examples. In the upper example a drilled hole helps to reduce the mass of metal on the left of the component, and the reader will observe the modifications to the corners. The mass of the gear shown below is reduced and made more uniform by machining it to an I section and, as cracks may develop from the blind holes in the end of the shaft, these are drilled right through.

The manipulation of components during heat-treatment plays dominant role in the success of the process. Heating should not be hurried and should be uniform from all directions. In a well-designed furnace, heat will be supplied equally from all sides; if flames are impinging on one portion of the part being heated while other portions are comparatively cool, trouble will ensue sooner or later. Sufficient time should be allowed for the heat to 'soak' right through the mass of the part, and this interval varies with its size. A 250 mm diameter bar would require about 3h at 900°C as compared with ¾ h for a bar 50 mm diameter. Given sufficient time during heating, a uniform structure can be obtained throughout the section of the component, but, when quenching takes place, the rate of cooling varies from outside to centre. The core of a large job after quenching may consist of ferrite and pearlite, even though the skin is composed of glass-hard martensite.

Table 2.5: Effect of mass on properties of quenched carbon steel 0.45% carbon steel water quenched from 870°C

Dia. of bar (mm)	Brinell hardness	Tensile strength (N/mm^2)	Elongation (%)	Redn in area (%)
17	550	1850	3	4
28	320	1032	12	28
76	241	800	19	47

During the process of quenching, the component should be lowered into the bath in such a way that it minimise distorting. For instance, if a tool such as a tap is quenched on its side, the sudden cooling and contraction of one side will promote bending; this is minimized by quenching vertically. Flat sections should be quenched edgewise and thick portions should enter the bath first. Once immerzed in the bath, the job should be moved about to prevent the formation of bubbles on the surface and consequent irregular disposal of heat. Water, one of the most efficient and drastic quenching media, provide maximum hardness, but it has the greatest risk of cracking and distortion. When extreme hardness can be sacrificed, oils such as whale oil, cotton-seed oil and the commercial quenching oils may be used. These give a slower rate of quench and the risk of quenching troubles is much less.

ALLOY STEELS

To provide material which will pass through a heat treatment process more successfully, many of the special alloy steel have been developed. By alloying small amounts of other elements, it is possible to influence the speed and temperature at which the hardening transformations occur, to influence and counteract the mass effect, to counteract the distortional influences, and so on. There are now available, for instance; alloy tool steels which reach their ideal condition of hardness by a relatively mild oil quench from a temperature of about 800°C. Furthermore, some of these steels possess a non-distortional characteristic which is necessary when making dies and other tools which cannot be ground after hardening. Examples of such steels will be found in various non-shrinking die steels which are available.

Properties given to steel by alloying elements

The information should help the reader when he encounters specifications giving the analyses of alloy steels, the applications of which are now so widespread.

Aluminium acts as a deoxidiser. It restricts grain growth and forms hard nitrides when heated in contact with nitrogen, resulting in extremely hard steel. Small amounts increase strength, but too much results in embrittlement. From 2% to 5% gives heat-resistance and resistance to oxidation.

Carbon is the essential constituent for hardening by heating and quenching. It increases the strength up to the saturation content of 0.87%, above which the steel becomes increasingly brittle unless other elements are present to take up the surplus carbon. It forms hard carbides with iron, chromium and vanadium, increasing the strength and wear-resistance.

Chromium froms hard carbides and gives deep hardening and good wear-resistance. Small amounts give toughness and increased strength and impact-resistance. It decreases machinability and decreases hardening range (temperature) unless balanced with nickel. It retains hardness at higher temperatures than iron carbide and gives slight red hardness.

Cobalt gives red hardness by retention of hard carbides at high temperatures. It tends to decarburise steel during heat-treatment. It increases hardness and strength, but too much decreases impact-resistance. It increases residual magnetism and coercive magnetic force in steel for magnets.

Lead forms minute strings and finely divided particles. Minute quantities give free machining without imparting the weakening effect of sulphur.

Manganese is a dioxidiser and desulphuriser. Small amounts increase hardness, strength and wear-resistance. It enhance solubility of carbon, lowers the critical point and widens the hardening range, thus permitting less drastic oil quenching. Air hardening begins at about 1.5%. At about 12% Mn, steel becomes austenitic at ordinary temperatures. High manganese increases the coefficient of expansion. Small amount increase depth and speed of hardening. It lower the tendency to distort during heat-treatment. Medium amounts produce brittleness unless other elements are present.

Molybdenum adds red hardness. It enhance strength and impact-resistance at high temperatures but hardens and embrittles at low temperatures. It retards grain growth at high temperatures, gives deep hardening, and widens the hardening range. It goes into solid solution but forms hard carbides when other elements are present. Small amounts reduce temper brittleness in certain steels. It increases the machinability of carbon steels.

Nickel enhance hardness, strength, ductility and impact-resistance. It narrows the hardening range but lowers the critical point, reducing danger of warpage and carcking. It refines structure and retards grain growth. It decreases machinability. It balances the intensive deep-hardening effects of chromium. Large amounts give resistance to oxidation at high temperatures.

Nitrogen is normally unwanted. It hardens slightly and decrease ductility. Small amounts refine grain and increase strength of high chromium steels. It forms hard nitrides with aluminium and is introduced externally into balanced Al steels. It promotes grain growth at high temperatures.

Phosphorus enhance cold-shortness. Small amounts increase strength slightly and increased resistance to corrosion.

Silicon is a deoxidiser and graphitiser. It throws carbon out of solution. Small amounts increase impact-resistance and strengthen low-alloy steels. Medium amounts increase magnetic permeability and decrease hysteresis loss. It forms hard iron silicides and large amounts give great hardness, wear-resistance and acid-resistance, but cause brittleness. It has a wide range of utility in alloy steels if used with expert technique.

Sulphur forms soft and weak sulphides which weaken the steel and promote hot shortness. Minute quantities promote free-cutting qualities.

Tantalum is used in some special steels to give increased resistance to scaling at high temperatures.

Tungsten adds red hardness by forming hard carbides which are stable at high temperatures. It widens the hardening range and gives deep hardening. It increases strength and wear-resistance. Small quantities produce fine grain structure, but large quantities tend to cause embrittlement. It forms hard, abrasive-resistant particles in high-carbon steels. It adds acid. and corrosion-resistance, and gives increased residual magnetism and greater coercive force in magnets.

Vanadium is a powerful deoxidiser. It toughens and strengthens steels. It forms hard carbides. It refines the grain and reduces grain growth. It widens the hardening range. Hardness is retained at higher temperatures than carbon steel. With chromium it froms double carbides, giving 'keen-edge' quality to steel.

DEVELOPMENT OF NON-FERROUS MATERIAL

These days, the service conditions imposed by the requirements of the aircraft industry, gas turbines, nuclear power, electronics, etc. have shown the need for properties in materials not always present in the more established range of ferrous and non-ferrous metals. A good example of this is given by the operation of the turbine blade of a modern aircraft jet engine. These blades rotating at the periphery of a large turbine disc at speeds up to 8000 rev/min are subjected to centrifugal forces of several tens of kilonewtons while operating at a bright-red heat. Such conditions are bad enough, but when we add to them the necessity of the freedom from failure associated with an aircraft and the results of a breakdown we realise that the requirements are stringent indeed.

Some of the materials being used previously belonged to the category of the 'rare' metals, but their properties and the improved processes for their manufacture are resulting in their more general use. We shall have space to give only brief general details of what is going on and advise the reader to pursue the subject in current specialized literature, for it is fairly certain that before long these metals will tools, the tool would provable last only for a minuts or so at a cuthing be commonplace in our work shops. At present some of them cause machining and other problems not experienced with the traditional metals we have been using for so long.

The *nimonic* series of chromium-nickel-base alloys possess high strength and corrosion-resistance at high temperatures. They are used for gast-urbine blades and for other applications where these properties are necessary. These alloys are very difficult to machine, prone to excessive work-hardening, and highly abrasive to the cutting tool.

Titanium is relatively light in weight, with a density of 4400kg/m^3 (Al 2800 and steel 7800 kg/m^3), but some of its alloys attain a strength of 925-1080 N/mm^2. It is found in wrought forms, such as plato, sheet and rod, and it can be drawn into wire. Its corrosion-resistance is high and it retains its strength at elevated temperatures. Titanium can be alloyed with many other metals to produce various alloys, and the chief alloying metals are aluminium, manganese, vanadium, iron, chromium and tin. Typical alloys prepared by ICI and Jessops are Ti-Al-Mn-ICI 314A or Jessops Hylite 40; Ti-Al-Sn ICI 317; Ti-Al-V ICI 318A or Jessops Hylite 50. The corrosion-resistance of titanium is bringing it into the chemical industry and its high strength-weight ratio into aircraft construction. The titanium alloys are extremely difficult to machine, having a pronounced tendency to gall up and cause chip weldment on the edge of the tool. With conventional tools, speeds are extremely low and tool life short; for example, to machine the 4% aluminium-manganese alloys with usual types of high-speed tools, the tool would probably lost only for a minute or so at a culthing speed of 10m/min. With grinding it is require to run the wheel at a peripheral speed about ¼ to ½ of the conventional speed. In all cases, machines of the maximum rigidity must be used for machining.

Uranium is a radioactive metal and it is used as a fuel in nuclear reactors. This metal is readily machined with high-speed-steel tools, but a copious supply of cutting fluid is necessary to reduce the dust hazard, minimise the risk of fire, and reduce tool pick-up. All machining must be carried out in special rooms with protection and monitoring for operators, special ventilation, and precautions against the ignition of the swarf.

Zirconium is a metal possessing low neutron absorption and is highly resistant to the corrosive action of water. It is being used for details of reactors and particularly for the cans to protect the fuel from the action of water coolants. It can be worked and forged and can be machined readily with high-speed tools. In the form of dust, swarf, etc. It is easily ignited, and the ignition temperature is low. This necessitates extreme precautions in handling and storing the swarf.

Beryllium. For canning and other type of purposes this metal is widely used. The best present-known method of producing it is by compacting from powder, and in its sintered compact form it can be rolled and extruded. Its most pronounced feature in the workshop is its largely toxicity, which enhances dangerous hazards to the lungs if dust is inhaled. For grinding and machining, therefore, extreme measures of safety are require, including the use of enclosed glove boxes for isolating the sources of danger. Beryllium must be cut with tungsten-carbide tools, owing to its abrasive character.

Niobium. Niobium has a melting point of 2468°C and, when suitably alloyed, promises to surpass the nimonic alloys in high strength temperature features. It is soft and ductile, being capable of much cold deformation. The mechanical properties of niobium are particularly susceptible to the content of oxygen at room temperature; for instance, with a content of O_2 of 0.16% it has a Vickers pyramid hardness of 194 and a tensile strength of 52.5N/mm^2 . As the oxygen content rises to 0.4% these Fig.s increase to 331 and 900 N/mm^2 respectively. Niobium has nuclear properties similar to zirconium and is used for nuclear canning, while its high heat-strength qualities are bringing it into the gas-turbine field. This metal is non-toxic and machines quite readily.

Magnesium. Magnesium is a silvery white metal with a relative density of 1.74 (aluminium 2.7). Owing to its extreme lightness it was used widely during the 1939-45 war for aircraft construction, and this resulted in a world production in 1943 of about ¼ million tonnes compared with 1000 tonnes in 1920 and 10 tonnes in 1900. Current world production is about 240, 000 tonnes. In its pure stats it has a tensile strength of about 110 N/mm^2, but this can be enhanced to about 310 N/mm^2 by alloying, the chief alloying metals being aluminium, zinc, manganese, zerconium and cerium. In its pure state it has a very large grain size, but by adding up to 0.7% of zirconium the crystals are reduced to the order of 0.1 mm, with corresponding improvements in physical properties. Magnesium alloys can be cast, rolled, forged, pressed and welded. It machines easily, but the swarf tends to be inflammable, although with suitable precautions the danger is not great. Magnesium Contain an extremely high strength-weight ratio and this is impressively brought home when, encountering a massive-looking component for the first time, we lift its weight having in mind the resistance that would be met with in any other metal. Magnesium occurs widely throughout the world and at present the most common method of producing it is by electrolysis from brine and sea water, which contains about 0.13%. It is also extracted by thermal reduction from magnesite.

ALLOY EAST IRON

In the properties of ordinary cast iron and to give qualities more suitable while retaining the important casting advantage, this metal process a large number of alloy cast iron have been developed to avoid certain inherent differences. These new irons are often quite unlike the conception we normally have of cast iron and in some cases their qualities are developed by heat-treatment after the normal process of casting.

Two recent examples of such materials are acicular cast

iron and spheroidal graphite (S.G.) cast iron. Acicular iron has nickel and molybdenum in its composition, has a tensile strength in the region of 460-620 N/mm^2 and is being used for cast internal-combustion-engine crankshafts and for components whose service is of a similar nature.

In S.G. iron the graphite content is converted from a flaky to a spheroidal form by the alloying of a small amount of magnesium. This change in the form of the graphite content of the iron raises the tensile strength to the order of 620-770-N/mm^2 and produces a tough metal which can be twisted and bent.

3

Engineering Materials

INTRODUCTION

Anything that are made up of matter, have weight and occupy space called materials. The earlymen started to use the material in the form of shelter. With the development of science and technology the demand for different kinds of materials started increasing. The end of the 19th century and begining of the 20th century gave birth of a new age of materials. During this time steel, aluminium, copper and plastics were developed and used extensively. The most exciting development of these days are polymers, ceramics and materials used in electronics and computers.

TYPES OF MATERIALS

Broadly there are three types of engineering materials as follows:

(1) Metals and alloys

(2) Glass and ceramics

(3) Organic polymers

Metals and alloys. These are also two types:

(a) Ferrous metals and alloys

(b) Non-ferrous metals and alloys

Ferrous materials contain iron as the main constituent. Some examples of ferrous materials are steel, wrought iron, cast iron, stainless steel and high speed steel.

Non-ferrous materials are those which don't contain iron as the main constituent.

Glass and ceramics. Glass is an inorganic polymer of high molecular weight. It is an amorphous solid and transparent mass. Various types of glasses used in industry and soda lime glass, lead glass, borosilicate glass and phosphate glass.

In engineering ceramics are basic oxides used for various applications. They have good thermal resistance and insulating properties. MgO, CdS, SiC, Al_2O_3, concrete, ferrites and garnets are commonly used for industrial purposes.

Organic polymers. Organic polymers are organic materials which have high molecular weight and high degree of plasticity. PVC, PTFE, polyethylene, fibre, cotton, nylon are the examples which are commonly used organic polymers in industry.

ALLOYS

These are the pure metals which are very rarely used for engineering applications because they are very soft and ductile. Most of the metallic materials used in engineering are combinations of two or more metals, knows as alloys. An alloy is any combination of elements, that results in a substance possessing good mechanical properties. Alloys are formed by a combination of different elements in different ways.

There is largest proportion in an alloy is known as the "base metal" while the other elements are known as "alloying elements". Some commonly used alloys in industry are cast iron (iron and carbon) steels (iron and carbon), wrought iron (iron and carbon), brass (copper and zinc), bronge (copper and tin), white metal (tin, lead and antimony), gun metal (copper, tin and zinc) and duralumin (aluminium and copper).

METAL CRYSTALS

The component atoms of metals on solidation assume an ordered arrangement relative to each other which completely satisfies the standard definition, "a crystal consists of periodically in three dimensions is a crystal.

THE FOURTEEN POSSIBBLE SPACE LATTICES

There are only fourteen possible different ways in which points can be arranged in space so that each point has indentical surroundings. Therefore, only fourteen different space lattices are possible as shown in Fig. 1. For many purposes as a whole, a space lattice is described as a unit cell. The unit cell of a space lattice is the smallest parallelogram having a lattice point at each corner which can be constructed within a given space lattice. Each face of a unit cell is therefore a parallelogram, with lattice points at all four corners.

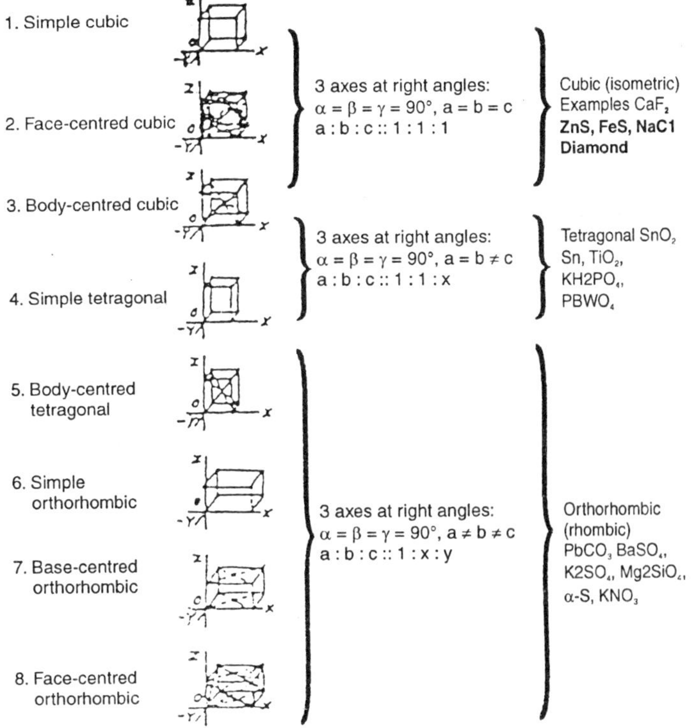

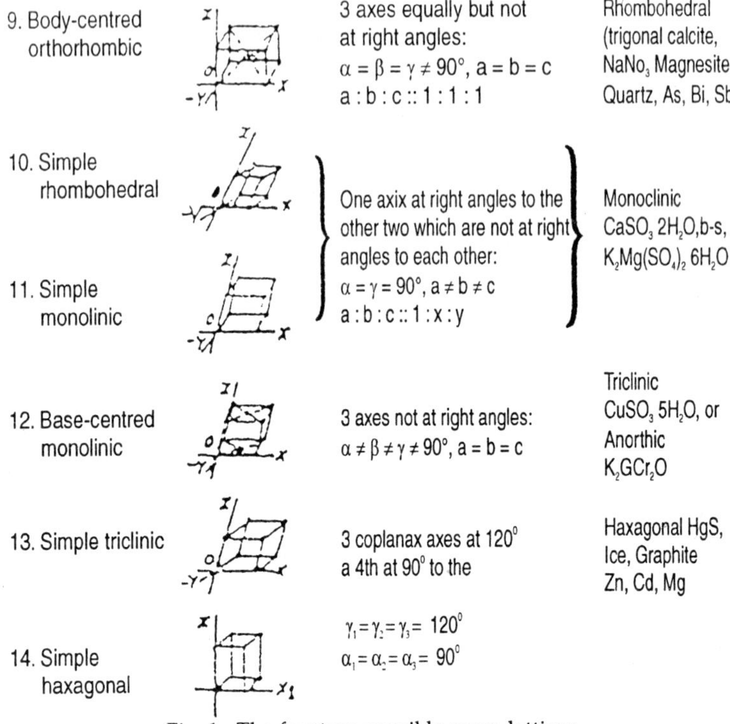

Fig. 1 : The fourteen possible space lattices

MOST COMMON CRYSTAL PATTERNS

In spite of only fourteen possible ways in which a crystal can rearrange itself in a systematic regular pattern, there are patterns are most common in metals.

1. FCC 2. BCC 3. CPH

FCC A face centred cubic structure can be properly understood by considering a unit cell as shown in Fig. 2. In metals crystallizing into face centred cubic pattern, an atom lies at the centre of each face of the cube and one atom at each corner of the cube. A face centred cubic structure is found in r-iron, aluminium, copper β-nickel, lead, gold palladium, β-rhodium, iridium, α-calcium, etc.

Table 3.1: Physical Properties of Pure Metals

S.No.	*Metal*	*Symbol*	*Colour*	*Melting Point°C*	*Boiling Point°C*	*Atomic weight*	*Density kg/m³*	*Specific heat*	*Coefficient of thermal conductivity X10–6sy 20°C*	*Electrical resistivity micro ohm*	*Thermal conductivity*
1.	Aluminium	Al	White having bluish tinge	659	1800	27.8	2700	0.215	23.0	2.68	0.507
2.	Copper	Cu	Yellowish red	1083	2310	63.5	8700	0.092	16.72	1.55	0.98
3.	Iron	Fe	Grey	1539	3010	56	7820	0.180	11.73	9.71	0.179
4.	Lead	Pb	Bluish grey	327	1620	207.19	11300	0.0307	29.12	20.60	0.081
5.	Magnesium	Mg	Silvery white	650	1100	24.3	1700	0.024	26.05	4.43	0.38
6.	Nickel	Ni	White	1452	2900	58.7	8900	0.105	12.75	6.847	0.151
7.	Silver	Ag	White	960	1952	107.87	10500	0.057	18.88	1.6	1.00
8.	Sodium	Na	Silvery white	97	880	23	972	0.037	–	1.3	0.07
9.	Tin	Sn	White with bluish tinge	232	2270	118.7	7300	0.054	21.4	12.81	0.163
10.	Zinc	Zn	Bluish white	419	917	65.4	7120	0.092	33.2	5.94	0.26

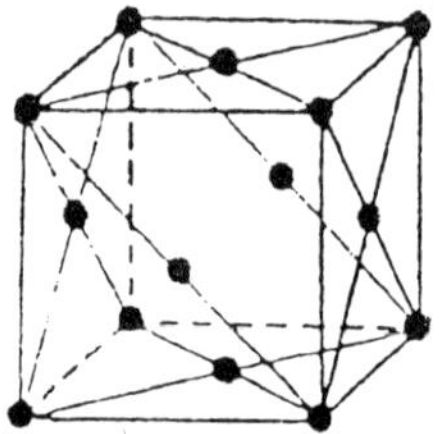

Fig. 3.2: Unit cell of face-centred cubic structure (FCC).

BCC The Unit cell of body centred cubic structure is shown in Fig. 3. In this system one atom lies at each corner of the cube and one stom at the body centre of the cube. The body centred cubic structure is found in α-iron, δ-iron, α-tungsten, molybdenum, vanadium, α-chromium, sodium, potassium, α-lithium, barium, tantalum, etc.

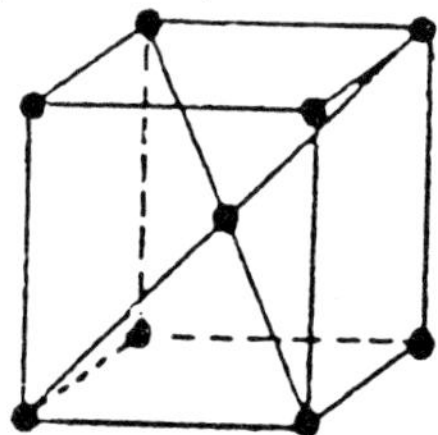

Fig. 3.3 : Unit cell of a body-centred cubic structure (BCC).

CPH In a closely packed hexagonal structure one atom lies at each corner of the hexagonal prism, one at the centre of each hexgonal face and three atoms are arranged in the middle as shown in Fig. 4. Magnesium, zinc, cadmium, berylium, α-nickel, β-chromium, α-cobalt, osmium, rhenium, molybdenum, technetium exhibit a closely packed hexagonal structure.

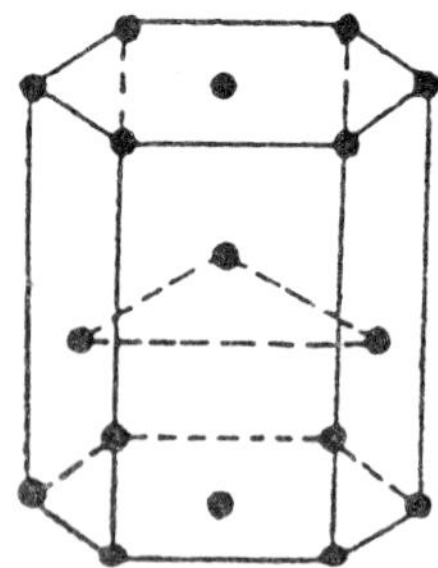

Fig. 3.4 : Unit cell of close packed hexagonal structure (CPH).

PIG IRON

Pig iron is the principal law material which is produced by smelting iron ore with coke and limestone in a blast furnace. The principal iron ores used for producing pig iron are as follows:

Haematite	Fe_2O_3	Red ore containing 70% iron
Magnetite	Fe_3O_4	Black ore containing 72.4% iron
Siderite	$FeCO_3$	Brown ore containing 48.3% iron
Limonite	Fe_2O_3-xHO_2	Brown ore containing 60% iron

The above-mentioned metallic are pure ores. Since most of the iron ores contain impurities, the metallic content is much lesser than that shown above. Haematite and magnetite are by far the most commonly used ores for extraction of pig iron.

The modern blast furnace plant for smelting pig iron consists of:

1. The blast furnace.
2. The conveyor belt system for carrying raw materials like ore, fuel and flux to the top of the furnace and charging them into the furnace.
3. The blower for sending the blast of air into the blast furnace.
4. Water pumping unit for supply of water to keep the walls of the blast furnace fool.
5. The appliances for cleaning the blast furnace gas.
6. A plant for generating power from the furnace gas.
7. A preheater for supply of heated air to the blast furnace.
8. Appliances for disposal of pig iron and slag.

The blast burnace has rearly 30 m height and its diamater is 8-10m in a vertical stock. It consists of an outer steel shell with suitable refractory lining. It is made up of two frustums of cones with their bases together. The upper frustum is called the "stack". It is supported by a mantle ring resting on 12 to 16 cast iron pillars all around the furnace. The lower frustum that appears to rest on the small or lower frustum is called the bosh.

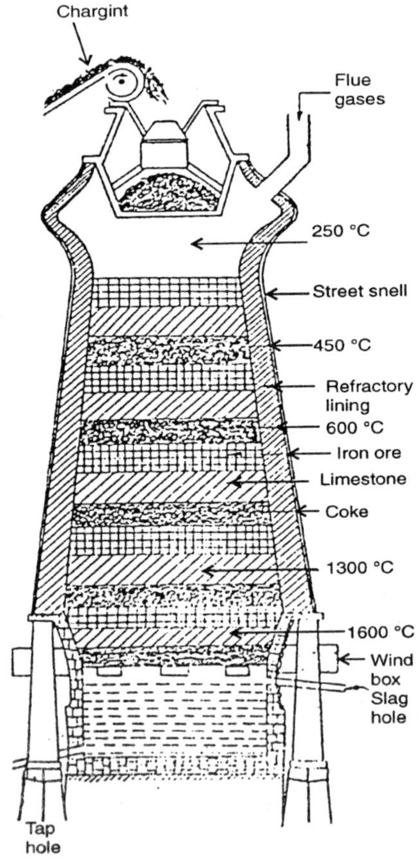

Fig. 3.5 3A.5 Section of a blast furnace.

Charging is done through the top of a stack, which gradually widens downwards to allow easy descent of the charge which expands on being heated due to the rising current of hot gases. The main chemical reactions taking place in the furnace are:

$2Fe_2O_3+6CO$	=	$4Fe + 6CO_4$
Fe_3O_4+C	=	$3FeO + CO$
$FeO + C$	=	$Fe + CO$
$CaCO_3$	=	$CaO + CO_2$
$CaO + SiO_2$	=	$CaO. SiO_2$ (Slag)

MODERN TRENDS IN BLAST FURNACE

Due to modern trends in blast furnace practice, some countries have succeeded in bringing down coke consumption from 800 kg to 500 kg per tonne of pig iron produced due to modern trends in blast furnace practice Some of the modern trends in blast furnace practice are:

1. By using a preheated blast.
2. Naphtha or other petroleum products may be injected along with the hot air blast.
3. Oxygen enrichment of the blast.
4. Humidification of the blast.

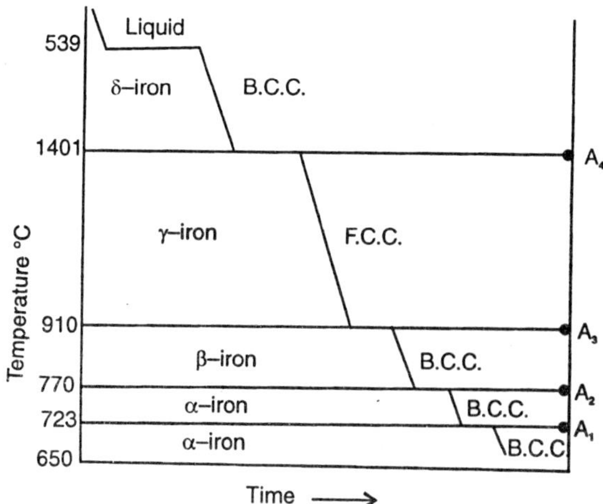

Fig. 3.6: Allotropic forms of iron.

ALLOTROPIC FORMS OF IRON

Allotropy means the phenomenon of existence of metals in various crystalline structures with changing temperatures and pressures.

When pure molten iron is allowed to cool, it shows various critical points at 1401°C, 910°C, 770°C and 723°C. These are known as $A_{4,}$ $A_{3,}$ $A_{2,}$ and A_1 points. At these temperatures arrests take place on heating and cooling and these are known as points of arrest. The arrest temperature (or change temperature) is slighty greater during heating rather than cooling. The temperature limit of 1539°C to 1401°C shows δ-iron having BCC structure. r-iron shows FCC structure and is stable between A_4 and A_3. β-iron is stable between 910°C and 770°C and shows BCC structure and α-iron is stable below 770°C but rearrangement of atoms takes place at 723°C. Below 723°C iron also has a BCC structure.

CLASSIFICATION OF IRON AND STEEL

We may clsssify iron and steel in different ways. Most commonly used methods of classification are (i) classification as per methods of manufacture and (ii) classification as per chemical composition. Since there are more than 1,000 types of iron and steel, mostly classification is done as per their chemical composition. In the market steel is supplied in different countries. USA, USSR, UK, and India have their own specifications.

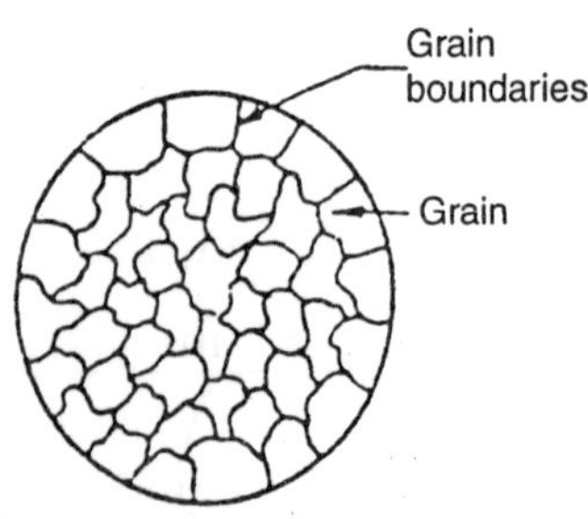

Fig. 3.7: Magnified microstructure of pure iron.

(i) Depending upon the method of manufacture, iron and steel and classified as follows:

(a) Pig iron a product of the blast furnace.

(b) Puddled iron (steel) manufactured by puddling process.

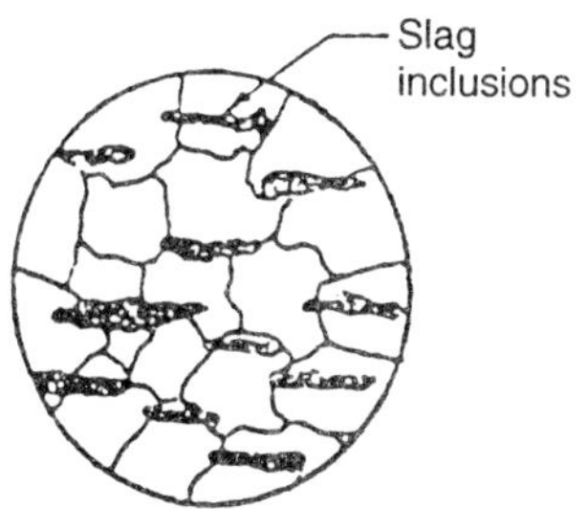

Fig. : Wrought iron (longitudinal section).

(c) Bessemer steel is the steel manufactured by Bessemer process.

(d) Open hearth steel is manufactured in an open hearth furnace.

(e) Electric are steel is manufactured in an electric arc furnace.

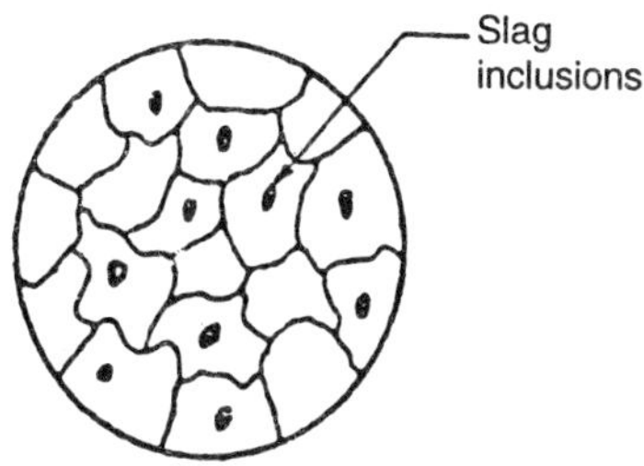

Fig. 3.9: Wrought iron (transverse section).

(f) Crucible iron is made in a crucible.

(g) Induction steel or steel manufactured by the induction process.

(h) Electrolytic iron is manufactured by electrolysis.

(ii) Depending upon chemical composition iron and steel can be classified as:

(a) Pig iron. It is manufacturec' by reduction or iron ore in a blast furnace and is the primary constituent of all types of iron and steel.

(b) Wrought iron. Wrought iron consists of high purity iron and iron silicate in its physical association. It contains a maximum of 0.01% carbon with slag distributed throughout the base metal in fibres as shown in Fig. 8 and 9 respectively. Wrought iron is produced by the oxidation of pig iron by removing carbon, silicon, manganese, sulphur and phosphorous. Pig iron contains nearly 6% of impurities but wrought iron is almost pure iron.

Propertions of wrought Iron. It is duclite when cold, and has good forming qualities. It has corrosium reaitaves better than steel. It can be readily worked and welded at temperatue close to its melting point. It is easily machinable and can hold protective metallic coatings. Mechanical properties of wrought iron depend largely upon the finished form and shape of the product after mechanical working. Wrought iron contains minimum amount of impurities and is soft having little strength and thus finds little use. Slag imparts corrosion resistance to wrought iron. Wrought iron is generally used in the form of sheets, billets, blooms and structural shapes.

Table : Uses of Various Steels

Carbon content %	*Various uses*
0.05 to 0.08	Highly ductile wires are manufactured.
0.08 to 0.15	It is also called rimmed steel which is used in the manufactures of sheets, strups, rods and wires when high suface finish is requrired, i.e. deep drawing, oil pans, etc.
0 15 to 0.20	Ship plates seamless boiler plates weldable boiler tubes camshafts, drum links, clutch fingers welded turbines.
0.20 to 0.35	Small forgings, crank pins, gear valves, crank shafts, railway axles, cross heads, connecting rods, rims for turbine gears armature shafts and fish plates.
0.35 to 0.45	Due to good machinability and deep hardening

	properties. These are in the manufacture of axles, special duty shafts, connecting rods, small and medium forgings cold upset wires and rods solid turbine rotors, rotor and gear shafts, shafts and brake levers.
0.45 to 0.55	This steel is useful for manufacturing articles, like railway coach axles, crank pins on heavy machines, large size forgings such as crank shaft stator, ring gears, axles spline shafts and hard drawn wire for tempered springs.
0.55 to 0.70	For production of drop forging dies, dies blocks, bolt heading dies, plate punches, set screws, self tapping screws, valve springs, cushion springs, lock washer springs, lock washer springs, thrust washers, etc.
0.7 to 0.8	These steels are used for making cold chisels, wrenches, jaws for vices, shear blades, hachsaw blades pneumatic drill bits, wheel for railwy service, automatic clutch disc, plough beams, etc.
0.8 to 0.9	Railway rail, plough shears, rock drills, circular saws, machine chisels, punches and dies, lock pins clutch discs, leaf springs, music wires, drawing dies, mover knives, etc.
0.90 to 1.00	Punches and dies, springs, balls of bearings, keys, pins, leaf and coil springs, harrow and seed discs.
1.0 to 1.1	Such types of steels are used for manufacturing railway springs, machine tools, mandrels, springs, taps etc.
1.1 to 1.2	Tapes, tools thread metal dies, twist drills, knives, etc.
1.2 to1.3	Files are manufactured.
1.3 to 1.5	Dies for wire drawing, paper knives and tools for turning chilled iron.
1.5 to 1.6	Saws for cutting steels and dies for wire drawing.

Uses of Wrought Iron

1. It is used for making iron bars for manufacture of stay bolts, rivets, engine bolts, etc. due to its corrosion and fatigure resistance properties.

2. Manufacture of plates.

3. Used in general forging applications.

4. Used in the manufacture of pipes.

5. Used in the manufacture of chains as it can be easily welded and possesses good impact strength.

(c) Cast iron. Cast iron has minimum of 2% carbon. Cast iron is brittle and not appreciably malleable at any temperature. It is generally manufactured from pig iron.

(d) White cast iron. It contains carbon in the combined form (Fe_3C) which makes it hard and brittle. It gives a white fracture and is therefore called white cast iron.

(e) Grey cast iron. Grey cast iron contains carbon in the free form called graphite. It gives grey fracture due to the presence of free graphite hence it is called grey cast iron. The names grey cast iron and white cast iron are devided from the colour of the fleshly fracture surface of the two materials.

(f) Malleable cast iron. When malleable cast iron is manufactured by a special heat treatment process of which cast iron, cementite splits into spheroidal graphite to produce malleable iron on heat treatment.

(g) Iron. Iron is pure iron which is soft, malleable and ductile.

(h) Steel. Steel is an alloy of iron and carbon and generally contains less than 2% carbon with substantial quantities of manganese and silicon.

Mild Steel

The alloys of iron and carbon containing carbon content from 0.08 per cent to 0.30 per cent. Mild steels cannot be hardened by heat treatment. These steels are having good machinability. Mild steel is used for making wires, nuts, bolts, nails, rivets, sheets, plates, tubes, rods, screws, structural steel sections, general purpose steel shafts, etc.

MEDIUM CARBON STEEL

It contains 0.3 to 0.6% carbon and is stronger than mild steel, possess better hardness and tensile strength but less ductility. These steels can be easily rolled, forged and welded. Their machinability is lesser than mild steel. Medium carbon steels are used for making stronger nuts, bolts, screws, axles, drop forgings, various steel sections, high tensile tubes, large forgings, dies, wires, ropes, hammers, agricultural components, locomotive rims, springs, etc.

HIGH CARBON STEELS

High carbon steels are which containing 0.6 to 2 per cent carbon. The strength of steel increases with increase of carbon and reaches maximum at 0.8 per cent carbon. Thereafter a little increase in hardness takes place but strength starts decreasing with increase of carbon content. Ductility and machinability of steel decreases with increase in carbon content. These steels respond fairly well to heat treatment and their mechanical properties can be altered appreciably with different heat treatment operations. Thus these steels are extensively used for making wood working tools, metal cutting tools, press working dies and punches, forging dies, small drills, small reamers hand files, springs, punches, etc.

ALLOY STEELS

New materials are needed for aircraft, atomic reactors, automobiles and space research. Plain carbon steels can meet these requirements partially. Alloying elements greatly enhance the specific properties and hence added frequently to steels. As per the Alloy Steels Reasearch Committee "Carbon steel are those steels containing not more than 0.5% manganese and 0.5% silicon." All other steels are regarded as alloy steels.

STAINLESS STEELS

Steels that don't develop stains easily and resist the action of corresion and oxidation are called stainless steels. Stainless steels are extensively used for making tabbware, cutlery, surgical and dental instruments, parts of water meters, ball bearings, pipelines, reaction vessels, stirring mechanisms and chemical plants.

Stainless steels can be classified as follows:

1. Plain chromium stainless steels
2. Chromium-nickel alloy stainless steels.

According to their microstructure, stainless steels have three groups as follows:

1. Hardenable or martensitic stainless steels.
2. Ferritic stainless steels.
3. Austenistic stainless steels.

Martensitic stainless steels These contain carbon from 0.10 to 0.80%, Mn from 0.5 to1.0%, silicon up to 0.4% maximum and chromium from 10 to 25%. Low chromium steels are used for making turbine blades, ball bearings and table cutlery. High chromium steels and used for making pump shafts, valve fittings and precision instruments. The composition range of stainless steels is shown in Table 3.3.

Table 3.3: Composition Range of Stainless Steel

Class	*C%*	*Cr%*	*Ni%*	*Uses*
Ferritic	0.1 to 0.25	16 to 30	–	Kitchenware, dairy components autom-obile fittings.
Martensitic	0.1 to 0.7	10 to 25	–	Turbine blades, ball bearings table cutlery.
Austenitic	0.08 to 0.25	15 to 25	5 to 25	Tableware, cutlery, chemical plants and ornamental goods.

Ferritic iron. Ferritic iron cantain 16 to 28% chromium, 0.1 to 0.2% carbon. 1.0 to 1.5% manganese. The chromium steels which have 25 to 30 have high resistance to attack by compound of sulphur and which have 16 to 18% chromoium steels have high resistance to corrosion, low impact strength and cannot be refined by heat treatment.

These are used for diary components, kitchenware, propeller shafts, aeroplane and automobile fittings and chemical plants.

Table 3.4: Heat Resisting Steel

Composition Percentage							
C	Ni	Cr	W	Mo	*Heat treatment °C*	*Structure*	*Uses*
0.6	–	10	–	0.5	0 Q 960, T 700	F+Cem	Exhaust valves of automobile,
0.78	1.5	20	0.5	–	0 Q1060, T 730	F+Cem	Aero and diesel engines. Good resistance to scaling.
0.25	0.5	15	–	–	0 Q 970, 730	F+Cem	Termocouple sheath retorts baffles in furnaces and has poor strength at high temperatures.
0.15	25	15	–	6.0	0 Q 1065,T 730	F+Cem	Gas turbine discs, tubes and blades.
0.10	12	16	0.5	20.0	0 Q 1040,T 720	A+Cem	–do–

Heat resisting steels. Steels that are capable of resisting the effect of heat and corrosion are known as heat resisatant steels. A material used at elevated temperature must possess the three properties as follows:

1. Resistance to oxidation and scaling.
2. Retention of strength at working temperature.
3. Structural stability at working temperature.

Commonly used head resisting steels are shown in Table 3.4. Composition Percentages

SILICON STEELS

Silicon forms iron silicide with iron. It is present in all types of steels up to 0.5%. In alloy steels the composition of silicon varies from 0.8% to 4.8%. Alloys of silicon are used in electrical machines as they have low hysterises loss. It reduces eddy current losses in cores of transformers and thus increases the magnetic permeability. Composition and uses of some typical silicon steels are shown in Table 3.5.

SPRING STEELS

The uses of spring steels are possess great strength, toughness and resilience. They must be capable of storing energy under a given load without permanent deformation.

Chemical composition and uses of some commonly used spring steels are mentioned Table 3.6.

HIGH SPEED STEELS

It was first produced in 1860 by Muschat. The actual forerunner of high speed-steel is developed in 1900 by Taylor's. In addition to tungsten, chromium and venadium were found to be useful due to their excellent wear resistance and hardenability. These steels retain a good cutting edge to a temperature of around 630°C. By this tool can be cut at speeds of 50-80 meters per minute with its nose at dull red temperature.

The ability of a tool to retain its cutting properties at high temperatures is known as red hardness.

Table 3.5: Composition and Uses of Silicon Steels

S No.	C	Si	Mn	Uses
		Percentages		
1.	0.1 max.	0.8 to 1.8,	0.1 to 1.3	Ductile, hence used for poles of DC instruments, armatures rotors and stators.
2.	0.5 to 0.6	1.0 to 2.7	0.8 to 1.0	High fatigue/strength, resillence and toughness. Used in quality sprins, rotors and stators of non-synchronous motors.
3.	0.5 to 0.6	2.7 to 3.7	0.7 to 1.0	—do—
4.	0.4 to 0.5	3.7 to 4.8	Mn 0.5 max. cr 6 to 10%	Good resistance to oxida- but poor strength high temperatures. Used in exhaust valves of motorcars, high speed diesel and petrol engines.

Table 3.6: Spring Steels

C%	Si%	Mn%	Cr%	*Uses*
0.6 to 1.0	0.2 to 0.5	0.6 to 1.0	—	Cheap, hence used in laminated springs for locomotives, wagons, carriages and heavy road rollers.
0.6	0.50 to	0.6	1.0	High quality and good elasticity, hence used in motor car laminated coils, coil springs and aircraft engine valves.
0.6	0.7 to 0.9	0.4 to 0.8	0.5 to 0.8	Leaf springs and helical spring for medium duty purpose.

Table 3.7 : Typical Compositions of High Speed Steels

C%	W%	Cr%	V%	Mo%	Co%	*Remarks*
0.65 to 0.67	14	4	1	–	–	Light duty
0.68	18	4	1	–	–	Normal duty
0.65	4	4	1	8.5	–	–do–
0.7 to 0.8	6	4	1	6	–	–do–
0.7 to 0.8	18	4	1	–	5	Fast cutting
0.65 to 0.75	18	4	2	1	10	-do-
0.65 to 0.75	20	4	1.5	–	12	-do-

High speed seels contain 18 to 19% tungsten, 3 to 5% chromium, 1.0 to 1.5% vanadium and around 0.6% carbon. Cobalt is often added to steels to increase hardness, hardenability and wear resistance. "Secondry hardness" is marked in such steels. Molybdenum acts as a cheaper substitue of tungsten. Typical compositions of commonly used high speed steels are shown in Table 3.7.

TOOL STEELS

Tool steels are classified on several factors like the operation to be performed, characteristics of material to be cut, machine tool to be used and rate of cutting. As per the society of automotive engineers these can be classified into the following six major groups:

1. Water hardening tool steels
2. Shock resistant tool steels
3. Cold working tool steels
4. Hot working tool steels
5. High speed steels
6. Special purpose tool steels.

The main constituents of water hardening tool steels are 0.7 to 1.5% carbon and 0.4 to 0.5% managanese. These are used for files, twist drills, clisels hammers, etc.

Shock resistant tool steel contains one or more alloying elements like manganese, chromium, tungsten, silicon and molybdenum. A commonly used shock resistant tool steel contains 0.5% carbon, 2% chromium and 0.5% tungsten. These steels are used for coal cutter picks, cold chisels, pneumatic chisels and punches.

The main alloying elements of cold working tool steels are manganese, tungsten and chromium. Table. 3.8 shows a few cold working steels alongwith their uses.

In hot working steels there are tungsten, molybdenum and chromium as alloying elements. Most commonly used hot working steel contains 0.3% carbon, 10% tungsten, 3% chromium, 0.3% molybdenum and 0.3% vanadium. It is used for hot drawing, hot forging and extrusion dies for casting aluminium, brass, zinc, and their alloys.

Special purpose tool steels contain a variety of alloys like nickel, tungsten, molybdenum, chromium and vanadium.

These steels are used for special purposes like stainless and heat resisting components.

Table 3.8: Cold Working Tool Steels

C%	Mn%	Cr%	W%	V%	*Uses*
1.0	0.95	0.75	0.5	0.2	It is also called oil hardening non-shrinking tool steel (OHNS). Used in master tools, gauges and dies.
1.1 to 1.5	0.6	0.75 to 1.5	4.0	0.3	It is called finishing tool steel. Used for twist drills, taps milling cutters, drawing dies, boring tools.
2.0	0.27	13.0	–	–	Oxidation resistant. Used in dies for blanking, coining, roller threading and drop forging of hard materials.

CAST IRONS

The basis of forming cast irons are identified due to natural result of the fact. The main disadvantages of cast irons are their brittleness associated with strength properties inferior to steels. However, cast iron is the cheapest of commercial alloys and is mainly used due to the following reasons:

1. It is the cheapest of all the alloys.
2. It has a low melting point and is easy to cast.
3. It is easily machinable.
4. It has good compressive strength.
5. It has good damping capacity.
6. It is relatively insensitive to presence of notches.
7. It is abrasion resistant.

TYPES OF CAST IRON

Cast iron manufactured in a cupola is relatively impure. It is generally produced in a cupola using coke as fuel and limestone as flux. Cast iron of higher purity is produced in an air furnace or electric furnaces. Since the electric furnace process is expensive, it is used mainly for producing malleable or nodular cast iron. Depending upon the type of furnace used, cast irons are classified as cupola irons, electric furnace cast irons or duplex irons.

On the basis of chemical composition and purity and their metallographic structures, these are classified as follows:

(a) Grey cast iron
(b) White cast iron
(c) Chilled cast iron
(d) Mottled cast iron
(e) Malleable cast iron
(f) Nodular cast iron
(g) Alloy cast iron
(h) Acicular iron.

Grey Cast Iron

Grey cast iron are the common cast irons of engineering and industry. Most of the carbon in grey cast irons is present in the form of flakes or graphite. The properties of grey cast iron dpends largely on the amount, size, shape and distribution of the graphite and combined forms of carbon. The metallurgist tries to control the effect of carbon largely by controlling the composition of carbon and silicon. Control of carbon depends upon the following variables:

1. Chemical composition of cast iron
2. Casting temperature.
3. Melting method used, i.e. cupola, duplex, air furnace, electric furnace, etc.
4. The rate of cooling from the the pouring temperature of the casting.
5. Heat treatment processes applied.

White Cast Iron

It is a mechanical mixture of ferrite and cementite. It is the distinctive appearance of its fracture which is clean and bright in colour resembling that of steel because of this reason it is called white cast iron. Most of the white cast irons are hypoeutectic in carbon content and generally contain less than 3% carbon. White cast irons can also be produced by lowering the silicon during melting and increasing the rate of cooling.

Table 3.9:Typical Composition of Cast Irons

Name	C%	Si%	Mn%	S%	P%
Grey CI	2.5/3.75	1.0/2.5	0.4/1.0	0.15	1.15
White CI	2.00/2.50	1.8/2.0	0.1/0.40	0.12	0.10
Malleable CI	2.20/3.60	0.4/1.0	0.1/0.40	0.03/0.3	0.10/0.20

NON-FERROUS METALS AND ALLOY

Non-ferrous metals and alloys are those contain a limited quantity of iron. Metals like copper, aluminium, magnesium, manganese, nickel, tungsten, chromium are frequently used for engineering purposes due to their following properties:

1. Good electrical and thermal conductivity.
2. Resistance to corrosion.
3. Good modulus of elastictity.
4. Good colour and colour imparting porperties.
5. Light metals like aluminium are used in aircarft manufacture.
6. They can be cast easily.
7. Good rolling, foring and machining qualities.
8. Good weldability.

On the basis of volume to volume, steels are much stronger than non-ferrous alloys, but on the basis of weight to weight aluminium and its alloys are stronger than steels.

copper

Copper is mainly found in USA, Mexico, USSR, China and Chile. In India, copper ores are found in Bihar, Orissa, Mysore and Sikkim. Pure copper prossesses good castability, weldability, machinability and workablity. See Table 3.1 for physical properties of copper.

Uses of Copper

Copper is chiefly used as a

1. Heat conducting material and
2. Electricity conducting material,
3. It is a corrosion resistant materal used extensively in boilers and heating vessels,
4. In electroplating and electrotyping.
5. For making coins, utensils and other alloys.
6. In manufacture of bearing materials.

Phosphor Bronze

It can be classifid into two main groups as follows:

Table 3.10: Alloys of Copper

Alloy	Percentage composition					Properties		Uses
	Cu	Zn	Sn	Pb	Al	Te		
Tellurium Copper	99.5	–	–	–	–	0.5	Good machinability	Electrical parts requiring high electrical conductivity.
Brass	70	30	–	–	–	–	Good ductility and strength	Cartridge cases, utensils.
Gilding brass	85	15	–	–	–	–	Golden in colour	Jewellery, compact boxes, etc.
Cap copper	98	2	–	–	–	–	Very ductile	Ammunition priming cap.
63/37 Brass	63	37	–	–	–	–	Ductile	Cold press work
60/40 (Muntz metal)	60	40	–	–	–	–	Properties of brass	Used as muntz metal.
Tin Bronze	88	2	10	–	–	–	General utility Structural bronze	Pressure castings. bushings, bearings.
Leaded tin Bronse	87	4	8	1	–	–	Corrosion resistant Up to 300°C	Steam pressure castings. bushing, gears, pumps.

High leaded tin bronze	Remainder	4	5 to 10	7 to 15	–	–	Used at high pressures and speeds, corrosion resistant	Bearing and corrosion resistant pumps.
Aluminium bronze	Remainder	–	–	–	9 to 11	Fe 1 to 4	Corrosion resistant having good strength at elevated temperature.	Heavy duty parts, marine equipment, gears bearings, bushes valve sets, guides, etc.
Leaded Nickel Bronze	Reminder	2 to 8	4 to 5	1.5 to 4.0	–	Ni 20 to 25	Corrosion resistant	Marine castings, hard ware, valves, dairy equipment.
Leaded red	Remainder	5 to 7	4to 5	5to 6	–	–	Corrosion resistant	Pipe and tube fittings plumbing goods, hard-ware, small gears impellers, pumps, etc.
Leaded Yellow brass	Remainder	24 to 36	1.0	1 to 3			Corrosion resistant	Ornaments, castings, hardware, valves, etc.
Gun metal	Remainder	2 to 5	5 to 10	–	–	–	Corrosion resistant having high tensile strength	Bearings, steam pipe fittings, marine com-ponents, hydraulic valves and gears.

(a) Cast phosphor bronze

(b) Wrought phosphor bronze

(a) *Cast phosphor bronze.* Its main constituenties are 5 to 13% phosphorus and remainer as copper. It is used in bearings, gear wheels, slide valves and gudgeon pins.

A 12% tin, 0.3% phosphorus bronze has a hardness of 100 BHN. It possesses good tensile strenght with 5% elongation.

(b) *Wrought phosphor bronze.* Its main constituents are 2.5 to 8.5% tin, 0.1 to 0.35% phosphorus and remainder as copper. It possesses high strength, good corrosion resistance and is mainly used as a spring.

Nickel Silver

Varying proportions of nickel are added to copperzinc alloys to increase their strength and utility. On the basis of the structure of materials there are two types of alloys.

(a) α-alloys

(b) α+ β-alloys

(a) *α-alloys.* Nickel has a remarkable decolorising effect on brass. Brass having 20% nickel appears silvery white. The alloys take a brilliant polish and have good resistance to corrosion. *α*-alloys contain 7 to 30% nickel with about 63% copper and remainder as zinc. These are called german silvers, but do not contain silver. These alloys can be cold, rolled and annealed. Annealing is done at a temperature of 650°C to 800°C according to their nickel content.

Nickel silvers are used in telephone and wireless industries for contactors and springs. laboratories and shops, spectacle parts, zip fasteners and latch keys. A 45% nickel alloy is used for parachute harnesses because of its high shear strength.

(b) *α + β-alloy.* It contains 45% nickel, 45% copper, and nearly 10% zinc. It can be extruded and is used for architectural and ornamental puroses because of its slivery white colour.

Inconel

The main constituents of Inconel are 80% nickel, 14% chromium, and 6% iron. It possesses good mechanical properties at elevated temperatures and can be cast rolled or cold drawn. It is mainly used for making springs and exhaust monifolds of aircraft engines.

Monel In monel there are 76% nickel, 30% Cu, 1.4% Fe, 1.0% Mn, 0.1% Si, 0.2% C. It has good resistance to sea and estuary waters, alkalies, reducing agents like sulphuric acid, oils containing brine and alkaline solutions. It is used in turbine blades, valve parts, pump rods, liners and impellers.

K-Monel In addition to above it, contains 3% aluminium and 0.5% titanium. Its mechanical properties are better than monel and is use in components requiring high strength with good corrosion resistance.

S-Monel In addition to all above, in contains 3 to 4% silicon. It also possesses good corrosion resistance and wear resistance. It is useed in components requiring high temperature and pressures like steam plants and oil tankers.

Nichrome The main constituents of nichrome are 65% Ni, 15% Cr and 20% Fe. It is heat and oxidation resistant and is used for making resistance wire of electrical furnaces and heating elements.

Permalloy Permalloy has 78.5% Ni, +21% Fe with traces of Cu, Co, Mn and Cr. It has high magnetic permeability and is used for making electrical impulses clear and good.

Alloys of Aluminium

Due to extreme lightness, alloys enable much greater volumes to be used for a given weight in aeroplanes, for high speed reciprocating parts, such as pistons and connecting rods. Aluminium alloys also have better balance, reduced friction and lower bearing loads. Chemical compositions and uses of commonly used aluminium alloys.

Duralumin

It is mostly used for general purposes.

Duralumin is extensively used for making forged, and stamped components like sheets, bars, rivets and tubes in automobile and aeronautical industries.

Table 3.11: Alloys of Aluminium

Alloy	Al%	Cu%	Mn%	Mg%	Ni%	*Uses*
Y-alloy	Rest	4.0	–	1.5	2.0	Used for mechanical working and castings.
Hindalium	Alloy of aluminium and magnesium					Trade name of Hindustan Aluminium Corporation Renukoot (Mirzapur) UP.
Magnalium	Remainder		–	2 to 10	–	Good physical properties and resistance to corrosion. Aircraqft and automobile components, dairy equipment and architectural parts.

BEARING MATERIALS

Due to the following qualities, bearing materials are very useful:

(1) Low coefficient of friction with the shaft.

(2) Good wear resistance, hard surface with tough core.

(3) Good thermal conductivity to dissipate the heat generated due to friction between the bearing and the rotating shaft.

(4) High compressive strength at operating temperature.

(5) Ability to retain the oil film at the bearing surface.

Table 3.13: Bearing Materials

S. No.	Alloy	Percentage composition					Uses
		Sn	Pb	Zn	Al	Cu	
1.	Copper base alloys	(a) 75	5	20	–	–	High bearing loads, locomotive slide valves and bearings for turn tables.
		(b) 80	10	10	–	–	-do-
2.	Aluminium base alloys	(a) 6.5	–	Ni	91	1.0	Corrosion resistant and good anti seizure properties. Better than babbits.
		(b) 6.5	Si 2.0	–	Balance		Used with high speed shafts.
3.	Tin base alloys or babbits	(a) 93		Sb 3.5	–	3.5	Good strength and toughness. Used as big end bearings for auto and aero engines.
	-do-	(b) 86	Sb 10.5	–	–	3.5	Hard and stronger than 3(a).Used in main bearing of auto and aero engines.
		(c) 83	4.0	Sb 10.5	–	2.5	Good shock resistance. Used as the main bearing.
4.	White metal (lead brass bearing)	20-25	63.5	Sb 1.5	–	1.5	Medium pressure and speed.

(6) Adequate plasticity under bearing load to distribute the load uniformly.

(7) Adequate strength at high temperature.

(8) It should not cause excessive wear of the shaft rotating in it, i.e. bearing material should be softer than the shaft material.

Table 3.12

Cu%	Mn	Mg%	Si%	Fe%	Zn%	Al%
3.5	0.60	0.45	0.58	0.53	0.58	Remainder
4.20	0.65	0.55	0.22	0.45	–	Remaineder

(9) Resistance to corrosion.

(10) It must be having small small pieces of comparatively hard metal embedded in a soft metal.

(11) It must have long life.

(12) It must be capable of bearing shocks and vibrations.

(13) It must have anti-seizure characteristics.

(14) It should be cheap and easily available.

ClassifiCation of Bearing Metals

There are three main groups of bearing metals in practice:

1. White metal type. consisting of hard particles embedded in a soft matrix.

2. Leaded bronze type, consisting of a hard matrix through which a softer metal is distributed.

3. Single metal or single phase alloy.

In addition to above properties of bearing materials, a very desirable property of a wide range of bearing alloys is that one of the constituents should possess a relatively low melting point. Taking into account the above properties of

Table 3.14: Properties and Uses of Rare Metals

S. No.	Element and symbol	Colour	Density kg/m^3	Melting point °C	Boiling point °C	Tensible strength N/mm^2	At. Wt.	Uses
1.	Cadmium (Cd)	White	8650	321	705	240-280	112.41	Cadmium base bearing alloys electroplating.
2.	Chromium (Cr)	White	7100	1850	2680	220-260	52.01	High speed steels, stainless steels, heat resistant steels, electroplating, pigments, chemicals and in chemicals and in chemical industry.
3.	Cobalt (Co)	White	8750	1492	2900	230 to 265	58.94	In mechanical properties it is next only to iron. Magnetic and retains magnetism up to 1100°C. Alloyed with steels. Used in permanent magnets.
4.	Iridium (Ir)	Grey	11300	327	1740	100 to 150	267.2	The rare and costly element, is used with platinum for ornaments and hard corrosion resistant alloys.
5.	Molybdenum (Mo)	Silvery white	10200	2625	5690	200 to 230	95.95	Filament wires, radio valves, filament supports and hooks.

Continued...

Table 3.14: Continued Properties and Uses of Rare Metals

S. No.	Element and symbol	Colour	Density kg/m^3	Melting point °C	Boiling point °C	Tensible strength N/mm^2	At. Wt.	Uses
6.	Nickel (Ni)	Lustrous white	8850	1452	2900	230 to 280	58.7	Alloying element in steel, stainless steel, heat resistant steel. As alloying element in monel, cupronickel, nickel silver food stuff containers, chemicals tanks, crucibles, compact bearings.
7.	Osmium (Os)	White	22500	3045	4600	175	190.2	Heaviest metal known. Not attacked by acids. Alloyed with iridium for thermometers, scientific apparatus, Jewellery, dentistry and chemical equipment.
8.	Platinum (Pt)	Greyish White	21500	1769	4100	160	195.23	Costly and rare element. Corrosion and acid resistant. Used in platinum thermometers scientific apparatus, jewellery, dentistry and chemical equipment.
9.	Rhodium (Rh)	Lustrous white	12400	1960	3560	165	102.9	Thermocouples, pyro meters,-extremely hard coatings. Corrosion resistant at high temperatures.

Table 3.15: Chemical Name, Trade Name and Uses of Plastics

S. No.	Chemical name	Trade name	Uses
1.	Polyethylene	Polythene, alkathene	Ice cube trays, developing trays, fabrics. collaposible nursing bottles.
2.	Polyvinyl chloride	Keroseal, tygon, geon	Plastic wood, water tubes, gramophone records.
3.	Polyvinyl acetate	Elvaset, Gelva	-do-
4.	Polyvinyl alcohol	Resistoflex, elvanol	Floor coverings, cheaper substitute for rubber.
5.	Polypropylene	Polyprex	Developing trays, water tubes, collapsible nursing bottles.
6.	Polytetrafluoroethylene (PTFE)	Teflon	Fountain pens, ping pong balls, kitchenwares.
7.	Cellulose acetate	Plasticele, lumerite	Films, window screens, gas masks, artificial silk.
8.	Cellulose nitrate	Pyralin, celluloid nitron	Fountain pens, ping pong balls, tooth-brush handles
9.	Polystyrene	Styrene	Battery boxes, radio parts, dishes, wall tubes, lenses.
10.	Phenol furfural	Durite	Instrument housings, brake linings.
11.	Urea formaldelyde	Plaskan, beatle	Tableware, light fixtures, instrument dials, clock cases.
12.	Polyster	Duraplex beckosol	-do-
13.	Phenol formaldehyde	Resimine, bakelite, catalin	Plugs, knobs dials, radio cabinets, knife handles, vacuum cleaner parts, electrical parts.

the main metal, bearing materials can be classified as

1. Babbit metals
2. Cadmium based alloys
3. Aluminium based alloys
4. Copper based alloys
5. Silver based alloys
6. Sintered bearing materials
7. Non-metallic bearing materials.

Other chemical composition properties and uses of metallic bearings is mentined. in Table 3.13.

Non-metallic Bearing Materials

There are two types of non-metallic bearing materials which commonly used are:

(a) Polytetrafluorethylene (PTFE)

(b) Nylon

Polytetrafluorethylene (PTFE) It can move without lubrication. Is extremely inert, having very low coefficient of friction (without lubrication), and low thermal conductivity. It is chemically inert to water, many chemicals and solvents and has good stability at high temperatures.

Nylon These bearings are used in limited quantity as they have a high coefficient of friction (0.14 to 0.32) as compared to PTE which is having (F.04). They have good resistance to corrosion but are used only at low loads and moderate or low speeds.

PLASTICS

Cellulose was the first component of plastics which was first produded in 18 56 by Schonbein. Later on another product named celluloid was invented by John Wesley in 1890. In 1924, an Austrian scientist developed urea formaldhyde, which was as transparent as glass. Plastics have been developed to such an extent that they are considered as very important engineering materials these days.

Plastics may be said as that material which contains an organic substance of large molecular weight as an essential ingredient, is solid in its finished state and at some stage of its manufacture can be shaped by flow. Plastics are superior to metals due to following characteristics:

1. They have good insulating properties.
2. Many plastics are transparent.
3. They possess good colouring properties.
4. They possess good surface finish.
5. Easy formation in different shapes is possible.
6. They possess good corrosion resistance.

Classification of plastics

Plastics can be broadly classified as

(i) Thermoplastic materials

(ii) Thermosetting materials

Thermoplastics

The materials which can be remelted to manufacture fresh products are called thermoplasts. Thermoplasts can be further classified into two types:

(a) Amorphous and (b) Crystalline.

(a) *Amorphous plastics*

(i) Polystyrene (ii) Methacrylate (iii) Polyvinyl chloride (PVC) (iv) ABS (Acrylonitrile, butadiene styrene) (v) Fluorinated polymers (vi) Cellulose (vii) Perpex.

(b) *Crystalline plastics*

(i) Polyethylene (ii) Polyamides (iii) Polyacetyl (iv) Polypropylene.

Thermosetting Plastics

These can be melted once and can't be remelted again are called thermosetting plastics.

They can be easily bent or shaped by applying moderate temperature and pressure. They are comparatively soft and possess less strength. They deform easily at 65°C to 85°C, Thermosetting plastics can be moulded only once but they possess better strength even at elevated temperatures. In the market, plastics are sold under various trade names. Thus a person must be familiar with the chemical name, trade name and uses of plastics.

PLASTICS POWER CONDUCTORS

In the GEC Laboratory UK, scientists first created a plastic in 1984 which was as good as electric conductor as metal. This process has opened a path to the world for electric and electronic components made of plastics. This phenomenon disapproved the general definition of plastics: "Plastics are light weight, pliable, easy to mould and excellent insulators." For these reasons, plastics have long been used to coat electrical wires and to make exterior parts of such common electrical equipment as telephones, power tools and computer terminals.

Regarding origination of plastic there is a story. in 1970, a Japanese chemist Hideki Shirakawa asked one of his students during class hour at the Tokyo. Institute of Technology to perform a routine experiment, "Turn a tiny amount of common accetylene welding gas into an organic polymer. The student, a Korean with limited Japanese vocabulary, misunderstood his professor and used far too much of a crucial chemical ingredient. Instead of getting the usual black powder the result in the experiment was a silvery film looking like a metal.

After five years, Alan McDiarmid, a University of Pennsylvania chemist, visited Japan. He was shown this silver like plastic. After that, research of the material started mainly due to efforts of McDiarmid and Shirakawa. Finally, it led to the discovery of plastics as electrical conductors.

For the purpose of taking advantage of this conductivity, industrial researchers have already produced scaled down experimental plastic batteries that weigh one-third of the conventional batteries.

It is important to note that plastic cells can be charged and discharged much faster than conventional bed acid batteries. In an electric car, such rapid delivery of power would vastly improve acceleration and hill climbing ability, both sorely deficient in today's electric vehicles. Plastics as conductors have more utility in video cassettes, recorders and solar panels.

Nylons were not existing 50 years ago but now these materials have replaced natural fibres in many applications. With the development of plastics as conductors, it may be said that these man made materials will find extensive use in the near future.

RUBBERS

About 75% of rubber is consumed solely by automobile tyre industry. There are three categories of rubber

1. Natural rubber
2. Synthetic rubber
3. Rubber like plastics.

Nutural rubber Is generally found in the countries situated in hot zone near equator, such as South African, Malaysia, Singapore, Mexico, Peru and Sir Lanka. Natural rubber is of botanical origin. It is found in the juice of many plants, like shrub quayule, Russian dandelion, mikweed and many other shrubs, vines and trees. The chief source of rubber is Hevea brassiliencis tree, that produces the best rubber latex. coagulated by acids or by a smoking operation, and the resulting spongy mixture is passed through rollers to form a sheet. This rubber is known as smoked rubber or crude rubber. The crude rubber is further treated by filters, plasticizers or softeners to produce commercial rubber.

Natural rubber possesses high strength, food resilience, abrasion resistance and low hysterises. it is readily attacked by organic solvents and disintegrates easily. It cannot withstand thermal changes.

Synthetic rubber Synthetic rubber is obtained by suitable combinations of selected monomers. These rubbers are based on models of natural rubber. Actually these are synthetic elastomers. Different types of synthetic rubbers are:

1. Styrene-butadiene rubber (SBR)
2. Butyl rubber
3. Nitrile rubber.

CEMENT

This is the most commonly used material in all types of concrete works. The commonly used cementing material is also called Portland cement. Portland cement is produced by partial fusion of a mixture containing limestone, clay, silica, alumina, iron oxide, and magnesium oxide.

Cement is manufactured these days in a variety of forms. Selection of cement for use in construction work depends on the

1. Type of structure
2. The strength of the structure desired
3. The conditions under which the construction is to take place.

Two mainly used processes for manufacture of cement are (a) dry process and (b) wet process. The important propertiesof cement can be classified under the following heads:

1. Rate of setting of cement.
2. Rate of hardening of cement.
3. Evolution of heat.
4. Resistance to abrasion and chemical action.

CONCRETE

Concrete is the most extensively used engineeing material. It is a building material obtained by mixing cement, aggregates and water in suitable proportions and then allowing it to set. The proportions of various ingredients used vary with the nature of work in which the concrete is to be used. Cement concrete possesses good compressive strength and is commonly used in the construction of columns, foundations, gravity dams, heavy foundations and bed blocks.

Commonly used aggregates in concrete can be divided into two main groups (a) fine aggregate and (b) coarse aggregate. The material that passes through 4.74 mm IS sieve size is termed as a fine aggregate. It should not contain more than 8 per cent of fine particles that must pass through IS sieve no. 15.

Concrete must possess sufficient strength without deterioration. It must be durable and workable. Good concrete must be able to resist wear and corrosion and should be water tight, compact economical and strong.

Composites Too often a single material can't fulfil the requirements of engineering. Thus combination of two or more materials is needed. The materals produced by combining two or more materials are known as composites. The various types of composites used in industry are

1. Whiskers
2. Carbon Composites
3. Glass fibre reinforced resin
4. Carbon fiber reinforced resin
5. Reinforced cement cement concrete
6. Metallic composites.

1. *Whiskers* are defect free crystalline materials, a few

microns in diameter containing $Al_2 O_2$, SiC and Si_3N_4 as the main constituents. Whiskers are produced by introducing these materials into resin and metallic materials. These are single crystals, free form dislocations. Another form of whisker contains silica fibre coated with aluminium. Whiskers were developed by Rolls Royce and are widely used these days in furnitures and utensils.

2. *Carbon Composite* was developed by R.A.E. Farnburough from commercially drawn acrylic fibres by subjecting them to various treatments and graphitizing to produce long threads. Two types of commercially used carbon composites are (a) high modulus carbon composite and (b) high strength carbon composite. Density of these composites is one-fourth of steel. On the weight to weight bases their strength is two times greater than steel.

3. *Glass fibres or resins* were first used in aeroplanes in World War II. Glass fibres possess good strength while the polymers have good toughness. The fibres are woven together and pressed into mats to form the composite. High temperature polyamide resin with pure silica fibres are used at high temperatures and possess good wear and fatigue resistance.

4. *Carbon fibre reinforced plastics* are produced from synthetic textile fibres, treated in such a manner that the side groups are totally removed. These composites possess properties similar to glass fibre reinforced resins. They possess lesser density, good strength and fatigue resistance.

5. *Reinforced cement concrete* combines the properties of tensile and compressive strength acting on structures. Steel possesses good tensile strength and concrete possesses good compressive strength. It is extensively used where both compressive and tensile loads are acting. Steel also possesses good bending properties.

6. *Metallic composites* These materials are main lyused to eliminate the disadvantage of brittleness and develop the

property of stiffness. Boron possesses low density, high modulus of elasticity but is brittle. It is used for reinforcing aluminium which is very ductile. This composite possesses good modulus of elasticity and toughness due to the presence of boron and aluminium. The composite is as stiff as steel but possesses only one-third of its density.

Metallic fibres like patented steel, stainless steel, tungsten and molybdenum wires are also used in a metal matrix of aluminium and titanium. These composites possess good strength and fracture toughness. Composites like sheet laminate promise interesting possibilities in the near future.

CERAMICS

The word ceramics has been derived from the Greek word "Keramos". It means potter's earth clay. The word ceramics is used to designate a large varietory of organic and inorganic material used at very high temperatures. Ceramics usually consist of nitrides, carbides, oxides, silicates, borates, etc. In ebgineering caramics include glass, abrasives enamels, non-metallic magnetic materials, high temperature refractories, etc.

Most ceramic materials contain silicates. Silicates are cheap, available in plenty and possess properties that make them useful engineering materials. Bricks, tiles, window glass, lead glass, protland cement contain a fair amount of silicates. Ceramics generally form covalen bond. Because all the valency electrons are occupied, ceramics are good insulators.

Classification of Ceramic Materials

Ceramics are classified in many ways. According of their common characteristic features, ceramics can be classified as

1. Clay products
2. Refractories
3. Glasses

Table 3.16 : Functional Classification of Ceramics

Group	Examples
Abrasives	Alumina, silica, Carborundum.
Fireclay products	Brickes, tiles, porcelein.
Pure oxide ceramics	MgO, $SiO_2Al_2\,O_3Cr_2O$
Cementing materials	Portland cement, lime and ordinary cement.
Rocks	Granite, sandstone, magnesite sillimanite, chromite.
Inorganic glasses	Coloured glass, window glass.
Minerals	Quartzite, felspar, calcite.
Refractories	Acidic, basic, neutral.

Depending on their industrial applications and structural criteria, ceramics can be classified in two ways:

(a) Functional classification

(b) Structural classification.

Table 3.17: Structural Classification of Ceramics

Group	*Examples*
Crystalline ceramics	Magnesite and alumina.
Non-crystalline ceramics	Neutral and synthetic glasses.
Glass bonded ceramics	Fire clay products.
Wood	Amorphous and crystalline.
Cement	Crystalline and non-crystalline.

GLASS

Glass may be said as an inorganic product of fusion, which has cooled to rigid condition without crystallizing. During cooling, glass doesn't change from liquid to solid state at a certain temperature like pure metals, but it remains in a non-crystalline vitreous state. In the beginning glass was considered a decorative material suitable only for window panes, bottles, pottery or art objects. But modern researches and methods of production have made it a material for engineering applications.

Classification of Glass

Glass is classified in many ways. According to use it is classified as optical glass, chemical glass, bottle glass, window pane glass, tableware glass, etc. According to chemical composition glass is classified as sodalime, lead glass fused silica, borosilicate, alumino silicate glass, etc.

Shaping of glass articles into various forms is usually accomplished by various casting techniques. Flat glass is produced by rolling a continuous stream of glass between water cooled rollers. Rods and tubes are made by a drawing process. Containers and special articles are made by pressing or blowing processes. A recent development is the production of a perfect surface by casting on molten tin. Most glass articles are now formed by highly complicated automatic equipments.

HEAT INSULATING MATERIALS

Materials used to prevent the loss of heat are known as heat insulating materials. Asbestos, cork, thermocole, asbestos cement sheeting, fibrelead and glasswool are common heat insulation materials.

Asbestos It is a heat resistant fibrous material used to describe six minerals. Chyrystolite, Amosite, Anthrophytlite and crocidolite are the most common. The most important deposits of asbestos are found in Canada. Asbestos is an inorganic material having thermal and electrical insulation properties. It is resistant to acids, chemicals oils and moisture and also possesses good strength.

Chrystolite Comprises nearly 93% of world's production and its main constituent is hydrated magnesium silicate. It possesses high tensile strength, good resistance to heat, excellent flexibility, resistance to weather and electricity. It is also alkali and acid resistant. It is mainly found in Canada and USSR and used in the manufacture of high temperature gaskets and packings, paper and boards, plastic paints and coatings.

Amosite Is found is South Africa. It is having comparatively low tensile strength and flexibility but good heat and chemical resistance.

Anthrophylite Is found in the USA and Finlana. It has low tensile strength but good acid resistance.

Crocidolite Is found in South Africa and Australia. It has high tensile strength, good resistance to acid and alkalies, but has poor heat resistane. It is used in the manufacture of asbestos cement sheets and chemically resistant packings and gaskets.

4

Metals and Alloys/Nonferrous

INTRODUCTION

The metals which do not contain iron as base called nonferrous metals. Melting points of nonferrous metals are generally lower than that of ferrous metals. Nonferrous metals suffer generally from hot-shortness, possess low strength at high temperature, and their shrinkage is generally more than that of errors metals. Nonferrous metals are useful for following purposes:

1. Resistance to corrosion.
2. Special electrical and magnetic properties.
3. Softness and facility of cold working.
4. Fusibility and ease of casting.
5. Good formability.
6. Low density.
7. Attractive colour.

Some of the principal nonferrous metals used in engineering work are: aluminium, copper, lead, tin, zinc, nickel, etc. and their alloys.

ALUMINIUM

This metal is prepared from a clayey mineral called

bauxite. It is a white coloured metal and produced by electrical process from the oxide (alumina). In India they are chiefly available in Bihar, Madhya Pradesh, Karnataka, Maharashtra and Tamilnadu.

Where a light noncorrosive metal is desired as in aircraft and automobile components where the saving of weight is an advantage, this silvery white used extensively. In its pure state the metal would be weak and soft for most purposes, but when mixed with small amounts of other metals it becomes hard and rigid. So it may be blanked, formed, drawn turned, cast, forged and die-cast. Its good electrical conductivity is an important property and aluminium is used for overhead cables. The high resistance to corrosion and its non-toxicity make it a useful metal for cooking utensils; under ordinary conditions, water and air have practically no effect on it. Since pure aluminium is more corosion resistant than any alloy of aluminium, a thin layout pure aluminium is generally clodded on the sheets of aliminium alloy to elimination corrosium. Again pure aluminium is not heat treataves. Allimunium alloy cladded by thin layer of aluminium can be heat treated. In this composite, the core containing aluminium alloy can be hardened. Roll bonding, a technique used for cladding sheets is shown in Fig. 4.1. Aluminum has the property of being beaten into foil and this aluminium foil is now widely used as silver foil. Aluminium metal of high purity have got high reflecting power and is, therefore, widely used for reflectors, mirror, and telescopes. The melting point of aluminium is 658°C and is having a specific gravity of 2.65.

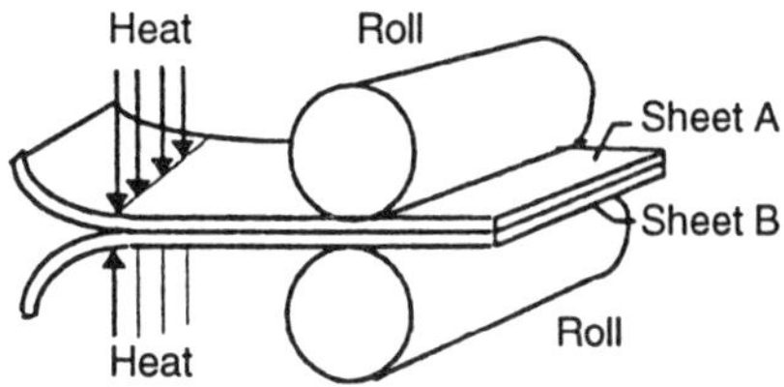

Fig. 4.1 Roll bonding method

Aluminium Alloys

The addition of small quantities of other alloying

elements converts this soft, weak metal into a hard and strong metal, while still retaining its light weight. Alloys can be classified as *cast* or *wrought,* both groups containing alloys that are age-hardened. The alloys in each of these two classes are further classified according to whether they respond to heat treatment of the strengthening type.

For general engineering use aluminium alloyed with small amounts of copper and zinc in the proportion of 12.5 to 14.5 per cent zinc and 2.5 to 3.0 per cent copper. An important series of casting and forging alloys having high strength have recently been developed for use in aeroplane construction. One example of such alloys is: zinc 5 per cent, magnesium 3 per cent, copper 2.2 per cent, nickel up to 1 per cent, aluminium the remainder.

An important and interesting wrought alloy is known as *duralumin.* This is composed of 3.5 to 4.5 per cent copper, 0.4 to 0.7 per cent manganese, 0.4 to 0.7 per cent magnesium, and aluminium the remainder. It is widely used in wrought condition for forging, stampings, bars, sheets, tubes and rivets. It is interesting because of its age-hardening property. After working if the metal is allowed to age for 3 or 4 days, it will be hardened. This phenomenon is called *age-hardening.* In the heat-treated and aged condition duralumin may have a tensile strength up to 40 kgf per mm^2 (400 newton per mm^2).

The *Y-alloy* contains 3.5 to 4.5 per cent copper, 1.8 to 2.3 per cent nickel, and 1.2 to 1.7 per cent magnesium. This alloy has the characteristic of retaining good strength at high temperatures. *Y-alloy* is, therefore, useful for piston and other components of aero-engines. It is also largely used in the form of sheets and strips, and after proper heat treatment this may be brought to minimum tensile strength of about 35 kgf per mm^2 (350 Newton per mm^2).

Berrylium copper are solution heat treated and precipitation hardened. Aging time is 2 to 3 hours at around 325°C. Aluminium bronzes containing more than 10 per cent aluminium are quenched from 650°C and subsequently tempered at a low temperature.

Table 5.1 Physical properties and uses of common aluminium alloys

Composition (Balance=Al) Strength	*Condition*	*Tensile **MPa*	*Yield (0.2% proof stress) MPa*	*% Elon gation 50mm gl*	*-BHN Number*	*Shear Strength MPa*	*Endurance *Limit MPa*	*Characteristic and uses*
1.2% Mn	A	110	41	30	28	75	48	A non-heat-treatable work-hardening wrought
	W	200	186	4	55	110	69	alloy used for general-sheet metal applications and cooking utensils.
4.0% Cu, 0.5% Mg	A	180	69	20	45	124	90	Duralumin, an age–hardening wrought alloy,
0.5% Mn, 0.5% Si	HT	425	275	20	105	262	125	hardened by quenching and ageing; used for aircraft constriction in the form of Alclad.
4.4% Cu, 0.8% Si	A	186	96	12	45	124	75	A strong age-hardening alloy for forging
0.8% Mn, 0.4% Mg	HT							and extrusion
7.0% Cu, 2.0% Si 1.7% Zn	As Cast	165	105	1.5	70	–	–	General purpose sand-casting alloy that is not

Continued...

Table 4.2: Physical properties and uses of common aluminium alloys

Composition (Balance=Al)	*Condition*	*Tensile Strength **MPa*	*Yield (0.2% proof stress) MPa*	*% Elon-gation 50mm gl*	*BHN Number*	*Shear Strength MPa*	*Endurance Limit MPa*	*Characteristic and uses*
								age-hardened and has low corrosion resdistance
4.0% Cu, 1.5% Mg 2.0% Ni	As	186	124	1.0	70	–	–	A sand-casting alloy capable of withstanding high temperatures.
12% Si	As Cast	270	145	2.7	–	–	–	A general purpose non-heat-treatable die-casting alloy having excellent corrosion resistance.
9.5% Si, 0.5% Mg	As Cast	300	186	3.0	–	–	–	A die-casting alloy having high strength and good corrosion resistance

* A=anneded, W=cold worled, HT=heat treated, G 500 million cycles in reverse bending.

COPPER AND ITS ALLOYS

Copper on account of its red colour, easily distinguished from all other metals. It is not found in pure state under the earth. Copper is extracted from copper ores, the chief being copper *pyrites.* Copper ore is first ground and then smelted in a reverberatory or small blast furnace producing an impure alloy called *matte.* In the second step air is blown through the molten metal to remove sulphur and iron contamination to obtain *blister* copper in the converter. In the last stage further refinement is carried out to produce pure copper (99.9 per cent) using electrolysis. For pyrites ($CuFeS_2$) and copper glance (Cu_2S), the roasting/smelting process liberates fumes of sulphur and arsenic which are destructive for environment. Collection of these poisonous gases is mandatory for the smelting unit. In India, copper ore is found at Ghatsila (Bihar).

Copper is relatively soft and is very mealleable, ductile and flexible, yet very tough and strong. A very efficient conductor of heat and electricity, being second only to silver, it is largely used in wire and sheet form for electrical purposes. Copper may be cast, forged, rolled and drawn into wires. It is non-corrosive under ordinary conditions and it resists weather very effectively. Copper in the form of tubes is used widely in mechanical engineering and also used in the making of munitions. The mechanical properties of copper depend upon its condition. Casting may have a tensile strength of 15 to 17 kgf per mm^2 (150 to 170 Newton per mm^2) which may be increased to 21.5 to 23.0 by working. The strength of hard-drawn copper wire may be as high as 38 to 46 kgf per $mm.^2$ The melting point of copper is 1,083°C.

Copper Alloys

Copper may be alloyed with a wide range of other elements to produce many different alloy groups of industrial importance. Some of the most important are:

1. Copper aluminium (the aluminium bronzes).
2. Copper-tin-antimony (babbitt metal).
3. Copper-tin (the tin bronzes).

4. Copper-tin-phosphorus (the phosphor bronzes).
5. Copper-zinc (the brasses).
6. Copper-nickel (the cupro-nickels).

Copper-Aluminium Alloy

Copper alloys contains 6 per cent and 10 per cent of aluminium and copper respectively. The aluminium gives the alloy lightness, while the addition of copper to pure aluminium increases its strength. The 6 per cent aluminium alloy has a fine gold colour, being used for imitation jewelary and decorative purposes.

LEAD AND ITS ALLOYS

Lead is extracted from lead ore called *galena,* which is a sulphide of lead. India is having very little lead ores. Lead has a bluish-grey colour and a dull, metallic lustre, but this is lost on exposure to the air, the surface becoming a dull grey.

Lead is a very soft, malleable and ductile metal and can be rolled easily. It is resistant to corrosion and many acids have no chemical action on it. Because of this it is used for water-pipes, roof covering, the sheathing of electric cables and for construction material of chemical plants. The melting point of lead is 327°C.

Lead Alloy

Alloyed with small percentage of arsenic, lead is used to produce shots for munitions. Lead alloyed with tin forms *solders* and alloyed with other metals makes bearing white metals and *type metal.*

TIN

In our country, the sources of tin are quite negligible. Tin is obtained from *tin stone,* and oxide by a refining process carried out in a reverberatory furnace. It is a brilliant white metal with a yellowish tinge. The melting point of tin is 232°C.

Soft, malleable and ductile, it can be rolled into very thin sheets. Tin does not corrode in wet and dry climates, making it useful as a protective coating for iron and steel. It is also used for tinning copper wire before the latter is made into cables.

BEARING METALS

There are following groups of bearing metals:

1. Copper-base bearing metals containing copper, tin and lead.
2. Tin-base bearing metals containing tin, antimony, and copper.
3. Lead-base bearing metals containing lead, tin, and antimony.
4. Cadmium-base bearing metals containing cadmium and nickel.

White-metal bearing alloys are those alloys in which tin, lead, and cadmium are predominating elements are designated. They have the properties of high plasticity combined with the low hardness and comparatively low melting point, which facilitates the formation of bearings by casting the metal directly in place and usually require no machining.

The *copper-base alloys* are harder and stronger than the white metals and are used for bearings which are required to resist heavier pressures. *Tin-base white metals* are used where bearings are subjected to high pressure and load, whereas for light loads and pressure *lead-base alloys* lead. Of the lead-base alloys, the alloys which contain 80 per cent lead and 20 per cent antimony is generally used. *Cadmium-base bearing metals* have more favourble properties, especially at elevated temperatures, than the tin-base alloys. The compressive strength of the cadmium-base bearing alloys are greater than those of the tin-base alloys:

For an efficient bearing combinations there should be the following conditions:-

1. That the shaft and bearing be dissimilar in their nature with bearing softer than the shaft.
2. That the most efficient bearing metal is one consisting of small pieces of comparatively hard metal embedded in the softer body of another metal.
3. That the bearing metals should have sufficient compressive strength to carry the bearing pressure, should wear to smooth surfaces as they rub together, and should develop a minimum of friction when they actually come in contact, as for example, when a shaft is starting or stopping.

BABBITT METAL

This is a tin-based white metal and contains 88 percent tin, 8 per cent antimony and 4 per cent copper. It is a soft material with a low coefficient of friction and has little strength. Babbitt metal makes a fine bearing and does not scour the shaft very easily when the lubricant fails. It is the most common bearing metal used with cast iron boxes when the bearings are subjected to high pressure and load.

COPPER-TIN ALLOY

The important copper tin alloys are—Bronzes, gun metal and bell metal.

Bronze

This is the composition in 5 to 25 percent tin and 75-95 per cent copper.

The alloy is comparatively hard, resist surface wear and can be cast into shape or rolled into wire, rods and sheets very easily. In corrosion resistant properties bronzes are superior to brasses.

Bronze is generally used in hydraulic fittings, pump linings, in making utensils, bearings, bushes, sheets, rods, wires, and many other stamped and drawn articles.

Phosphor Bronze

There are some kinds of bronze as follows Phosphorus bronze is the compositions varies according to weather it is to be forged and wrought or weather made into castings. A common type of wrought phosphor bronze has copper 93.7, tin 6, and phosphorus 0.3 per cent.

Phosphorus increases the strength, ductility, and soundness of castings. The alloy possesses good wearing quality and high elasticity. The metal is resistant to salt water corrosion.

Pump parts, linings, and propellers are examples of cast manufacture. A variety of phosphor bronze suitable for casting contains 11 per cent tin and 0.3 per cent phosphorus alloyed with copper. This is used for bearings which must carry heavy loads, worm wheels, gears, nuts for machine lead screws, springs and many other purposes.

Silicon Bronze

This contains 96 per cent copper, 3 per cent silicon, and 1 per cent zinc. Silicon bronze has the good general corrosion resistance of copper, combined with higher strength, and in addition can be cast, rolled, stamped, forged, and pressed either hot, or cold and can be welded by all the usual methods.

It is used in parts for boilers, tanks, stoves, or wherever high strength and good corrosion resistance are required.

Manganese Bronze

Manganese bronze is an alloy of copper, zinc, lead and a little percentage of manganese. The metal is highly resistant to corrosion. It is stronger and harder than phosphor bronze.

This is generally used for preparing bushes, plungers and feed pumps, rods, etc. Worm gears are frequently made from this bronze.

Gun Metal

This contains 10 per cent tin, 88 per cent copper, and 2 per cent zinc. The zinc is added to clean the metal and increase its fluidity.

It may be forged when at about 600°C. The metal is very strong and resistant to corrosion by water and atmosphere. Originally, it was made for casting boiler fittings, bushes, bearings, glands, etc.

Bell Metal

Bell metal contains 20 per cent tin and the rest is copper. It is hard and resistant to surface wear. Bell metal is used for making bells, gongs, utensils, etc.

ZINC AND ITS ALLOYS

In the extraction of the metals, the ore is first roasted in a reverberatory furnace to convert the sulphide to oxide, and in the case of calamine, to drive off carbonic acid and water. The roasted ore is then reduced either in a furnace or electrolytically. Zinc is a fairly heavy, bluish-white metal used principally because of its low cost, corrosion resistance, and alloying properties. The melting point of zinc is 419°C.

The oldest and most important methods of applying the zinc coating are known as *galvanizing.* When rolled into sheets, zinc is used for roof covering and for providing a damp-proof noncorrosive lining to containers, etc.

Zinc casts well and forms the base of various die-casting alloys. A typical high strength die-casting alloy would have the following composition and properties:

Table 4.2: Typical die-casting alloys

Composition		
Cu= 1.25 per cent	Al= 4.0 per cent	Mg=0.08 per cent
Fe= 0.10 per cent	Pb=0.007 per cent	Cd=0.0007 per cent
Sn=0.007 per cent	Zn=remainder	
	Properties	
UTS = 32 kgf/mm^2 or 320 Mpa	per cent elongation= 7 per cent	
Charpy Impact No. = 65 J or 6.5 kgm		Melting point = 425°C

COPPER-ZINC ALLOY

The copper-zinc alloys are mostly used in brass and muntz metal.

Brasses

Brasses are fundamentally a binary alloy of copper with as much as 50 per cent zinc. Various classes of brass, depending on the proportion of copper and zinc, are available for various uses. Suitable types of brass lend themselves to the following processes. casting, hot forging, cold forging, cold rolling into sheets, drawing into wire and being extruded through dies to give special shaped bars. The melting point of brass ranges from 800°C to 1,000°C.

The alloy is noncorrosive and air, water and some acids do not appreciably affect it. It is soft, ductile and has tensile strength with good fusibility and surface-finish characteristic. It is non-magnetic and is poor conductor of electricity. By adding small quantities of other elements the properties of brass may be greatly changed, as for example, the additions of 1 or 2 per cent of lead improves the machining quality of brass. Small amounts of tin are sometimes added to brass to increase its hardness.

Brasses are used in hydraulic fittings, pump, linings, in making utensils, bearings, bushes, etc.

Table 4.3: Composition, properties and uses of brasses

Name	*Composition* *Copper*	*Zinc*	*Other elements*	*Microstructure*	*UTS, depending upon working (MPa)*	*Uses*
Copper	100	–	–	Pure metal	150-480	Tubing, piping, sheet, wire, electrical conductors
Guilding metal	90	10	–	Alpha	280-505	Forgings, rivets, jewellery applications.
Tombac or French Gold Low brass	80	20	–	Alpha	325-825	Drawing and forging operations.
Cartridge or spinning brass	70	30	–	Alpha	365-635	Cartridge cases, condensor tube, sheet fabrication, a general purpose brass.
Admiralty brass	70	29	Sn=1	Alpha	365-635	Condensor tubes exposed to salt water (high corrosion resistance)
High metal	66	34	–	Alpha	365--860	Stamping and drawing operations.
Muntz metal	60	40	–	Alpha-beta	410-860	Suitable for many hot working operations; rolled also cast valves and marine fittings.

Continued...

Table 4.3 (*Contd.*)

Name	*Composition* *Copper*	*Zinc*	*Other elements*	*Microstructure*	*UTS, depending upon working (MPa)*	*Uses*
Naval brass	60	39	Sn=1	Alpha-beta	610-860	As above, but possess increased corrosion resistance.
Tobin "bronze"	60	38	Sn=1 Al=1	Alpha-beta	410-860	Brazing alloy for naval brasses, etc.
Manganese "bronze"	62	32	Al=4 Fe=1.5 Mn 2.25		410-1035	High tensile casting metal, yield point =510 MN/m^2, UTS=710 MN/m^2.
Brazing brass	50	50	–	Beta		Brazing rods.

Muntz Metal

Mentz metal contains 40 per cent zinc and 60 per cent copper. In rare cases lead in small quantity also included. It is stronger, harder, and more ductile than common brass. It is excellently suited for working at 700^{0}C but not for cold working. Muntz metal is used for a wide variety of small components of machine, electrical equipment, fuses, oldnames works, and for bolts, rods, tubes, etc. Is widely employed in making such articles which are to resist wear. Table .53 gives the properties and uses of some of the important brasses.

NICKEL AND ITS ALLOYS

All nickel production is obtained from sulphur are, Although some nickel is obtained commercially from oxide ores, arsenical ores, and the ores of copper, manganese and iron, at least 85 per cent of all nickel production is obtained from *sulphide ores.*

The from of pure nickel is tough, silver-coloured metal, rather harder than copper, and of about the same strength, but possessing somewhat less ductility. It closely resembles iron in several of its properties, being malleable and weldable, and perceptibly magnetic, but unlike iron it is little affected by dilute acids, is far less readily oxidisable, and deteriorates much less rapidly under atmospheric influences. For this reason articles of iron and steel are frequently nickel-plated to protect them from rusting. Nickel is much used for cooking utensils, and other vessels for heating and boiling. Nickel enters as a constituent into a large number of ferrous and nonferrous alloys, and frequently finds application as a catalyst in important industrial processes.

The cases of nickel alloys are as follows:

Monel Metal

Monel metal contains 60 per cent nickel, 38 per cent copper and aluminium or manganese in small quantity.

This is white, tough, and ductile metal and can be readily machined. It welds without difficulty. It can be heat-treated.

It is also resistant to corrosion by most agents and has high strength at elevated temperatures. Monel metal is used in the forms of rod, sheet, wire, and welded tubing. It is widely employed for structural and machine parts which must have a very high resistance to corrosion and high strength as steam turbine blade, impeller of centrifugal pump, etc.

German Silver

German silver contains copper 25-50 per cent, nickel 10-35 per cent and zinc 25-35 pre cent. Tin and lead are also included in some cases.

German silver is hard, white and ductile. This is usually of bright silvery colour, but may also assume various other pleasing colours by adjusting the proportions of copper, nickel and zinc. It has in addition to colour, good mechanical and corrosion resisting properties. It is also known Ni-silver.

For making utensils and resistances in electrical work, it is used.

Inconel

Inconel contains 75 to 80 per cent nickel, 10-15 per cent chromium and the rest iron. It can be used for parts that are exposed to high temperature for extended period. It is less reactive than nichrome to acid.

Nichrome

Nichrome is an alloy of nickel with chromium and is used widely as resistance wire for electrical appliances.

Nimonics

This one of the new type of nickel which is being developed by proper heat treatment attain excellent properties for very high temperature service as well as under intermittent heating and cooling conditions. They contain 15 to 18 per cent chromium, 15 to 18 per cent cobalt, 3.5 to 5 per cent molybdenum, 1.2 to 4.0 per cent titanium, 1.2 to 5.0 per cent aluminium, and the remainder nickel.

ALLOYS FOR HIGH TEMPERATURE SERVICE

This has made to develop a number or highly specialized alloys. Nickel or cobalt forms the base metals in this range of alloys. Most of these alloys possess yield strength in excess of 70 MN/mm^2 or 700 kgt/mm^{2}and 250 to 370 Brinell hardness number at room temperature. Table 4.4 shows some typical-high-temperature alloys.

METALS FOR NUCLEAR ENERGY

Metals, demand and has been created by nuclear projects to withstand stringent condition imposed, and as a result some metals, previously considered "rare" are now coming in the use in a wide perspective. The various metals for producing nuclear energy are used as raw materials, moderators reflectors, fuel elements, fuel canning materials, control elements, and pressure vessel materials. Thus, uranium, thorium, plutonium, zirconium, beryllium, niobium and their alloys are primarily used for nuclear engineering purposes.

Table 4.4: Alloys for high temperature service

			Types of alloy		
Composition	*Nimonic 80A*	*Inconel 713C*	*Incoloy 910*	*Hastelloy*	*Vitallium*
Nickel	Balance	Balance	42.0	45.0	2.5
Chromium	21.0	12.0	13.0	22.0	28.0
Cobalt	–	–	–	1.5	62.0
Tungsten	–	–	–	0.5	–
Titanium	2.5	0.5	2.4	–	–
Iron	–	–	Balance	Balance	1.7
Tantalum	–	2.0	–	–	–
Aluminium	1.2	6.0	–	–	–
Carbon	0.04	4.5	0.04	0.15	0.28
Molybdenum	–	–	6.0	9.0	5.5

Uranium

Uranium is used for nuclear energy production. This is used as a nuclear fuel and is radioactive, easily oxidized and exists in three allotropic forms. This has a poor resistance to corrosion and needs to be protected for use as fuel elements by roll cladding a thin aluminium or zirconium jacket.

The metal in the pure condition is weak and is susceptible to severe irradiation damage and growth in the reactor environments. Addition of some alloying elements such as chromium, molybdenum, plutonium, ziroconium, etc. are added to make the material highly suitable for nuclear power. Uranium compounds, such as UO_2 as a dispersion is cermets or as ceramic slugs, have been found to give better service.

Uranium oxide is highly refractory, shows no phase change in an inert atmosphere, possesses a good strength and a high corrosion resistance. But it has a low thermal shock resistance, poor thermal conductivity and high coefficient of expansion. The melting point of uranium is 1850°C.

Thorium

Thorium is used as a fuel which is free from phase changes below 1480°C. Thorium ($_{90}Th^{234}$) is also a radioactive metal like uranium and this can be converted into uranium ($_{92}U^{234}$) by beta decay. This mechanical properties of thorium which is soft and weak when pure, are drastically changed by small addition of impurities. Only 0.2 per cent of carbon raises its tensile strength from 14 to 38 kgf/mm^2 (140 to 390 MPa). Small additions of titanium, zirconium, and niobium decreases the strength and hardness of the metal. Uranium addition increases the strength of thorium.

This is an emitter of alpha rays and releases considerable quantity of the radioactive products during processing, but this being a cubic metal (*fcc*) is less susceptible to irradiation damage. The melting point of thorium is 1845°C.

Plutonium

It is produced through neutron absorption by fertile U^{238} and subsequent beta decays. Plutonium melts at 640°C. It is extremely toxic and emits alpha rays. Due to its being naturally fissionable, plutonium continues to emit high energy gamma and other radiation.

The metal is chemically more reactive than uranium and has poor resistance to corrosion. It has six allotropic forms.

The chief application of plutonium is in the production of atomic weapons and also in breeder reactors. Since some plutonium alloys and compounds emit neutrons, adequate health protection is necessary while using these materials.

Zirconium

Zirconium contains 0.5 to 2 per cent hafnium which is a strong absorber of neutrons and must therefore be removed. The main use of zirconium is for cladding fuel elements and for structural components in water-cooled systems. So it must have increased corrosion resistance. Zircaloy-2 containing 1.5% Zn, 0.1% Fe, 0.05 % Ni, and 0.1% Cr provides better corrosion resistance and is generally used in water-cooled reactors.

Zirconium has a relatively poor resistance to CO_2 at elevated temperature, but this is improved by the addition of 0.5 % Cu, 0.5 % Mo with an increase of tensile strength to 51 kgf/mm^2 (510 MPa) and improved creep resistance at 450°C. This zirconium is specially useful in gas cooled reactors. The melting point of Zirconium is 1750°C.

Beryllium

The meting point of Beryllium is 1280°C. The metal is used as a moderator, reflector and neutron source. This is very reactive and forms compounds with the furnace atmospheres and refractories. Vacuum or inert gas are necessary during melting. The cast metal is usually coarse grained and brittle, and powder metallurgy methods are employed in the fabrication of beryllium components.

Beryllium is a toxic metal. Its compounds are also very toxic. Safety measures are to be taken when handling this metal.

Niobium

The melting point of niobium is 1950°C and good strength, ductility and corrosion resistance especially to liquid sodium coolants, excellent capacity of coexistence with uranium. Its oxidation resistance above 400^0C is indifferent, but is greatly improved by alloying. The metal niolium titanium hetrium, tantalium and zireonium alloys are al used in set aircraft, reactors, missiles and for nuclear reactor.

5

Carpentry Shop and Wood Working Tools

INTRODUCTION

Any class of work with wood is called carpentry. The all works like roofs, floors, furniture, etc. are come under carpentry, while making of doors, windows, cupboards, dressers, stairs, and all the interior fitments are come under joinery.

Timber is the basic material used for any class of wood working. The term 'timber' is applied to the trees which provide us with wood. Wood is one of the most valuable biodegradable raw materials of industry and daily uses. It is available in a wide choice of weights, strength, colours and textures. Wood can be sliced, bent, planed, sawed and sanded due to its good characteristics.

WOOD'S STRUCTURE

Inside the trees due to the fact that each year a new layer of tissue is formed on the outside of previous layers. The layers are termed *annual rings,* because most trees produce one ring each year. Each annual ring is composed of an open porous layer known as springwood, due to the rapid growth in spring, and a thinner denser layer called autumn wood, due to the slowing down of the growth in late summer and autumn. In spring the outer rings of the tree known as *sapwood* convey the watery sap from the roots up to the leaves. Here it undergoes certain chemical changes, and on its return journey to the roots in autumn, this perfected sap leaves behind various starchy secretions, gums, or resins, according to the type of tree. These substances fill up the tissues, feed the tree, and help to form a denser wood known as Doormen or *heartwood*. The heartwood is dead as far as the growth of the tree is concerned. The cambium layer situated between the sapwood and the bark is responsible for the formation of new wood each year. It divides up, forming a layer of new wood cells on the inside and a soft layer on the outside which becomes bar. Each year the innermost layer of sapwood becomes transformed into heartwood.

It is observed that heartwood is the only part of the tree which should be converted for use, as sapwood is more prone to attack from wood-destroying organisms. If sapwood is properly treated with an effective preservative, it is as durable as heartwood, similarly treated, when used under conditions favourable to decay. The difference in colour between sapwood and heartwood is most important when choosing wood for joinery, the decorative value of which depends on the dark colour and grain of the heartwood. Treating with preservatives is useless and staining is very difficult, as sapwood absorbs stain more readily than heartwood and takes on a much darker shade. Joinery work is concerned with carpentry.

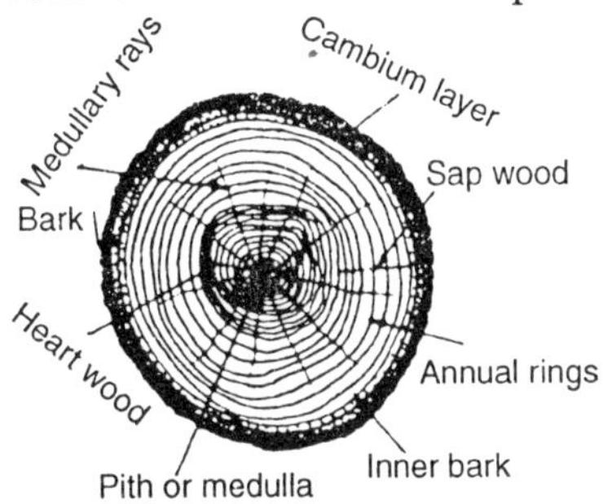

Fig. 5.1: Cross section of a log.

THE GRAIN OF THE WOOD

Appearance of the wood on any of its cut surface refers

to wood grain. It is due to variations of ring growth and of colour in the wood. There are two types of wood on the basis of variations of grain:

(i) Open-grained wood. An *open-grained wood* such as oak has minute pores over its exposed surface. In standard wood-finishing methods, these pores are levelled with a coat of filler.

(ii) Close-grained wood. A *close-grained wood* such as fir or pine has no such pores in its surface.

To make plane surface the grain of the wood also refers to the direction of the cellular or fibrous structure of the wood, which is the longitudinal direction. Timbers for structural use must be so cut that the grain runs parallel to the length of the timber; otherwise there is a marked reduction in strength.

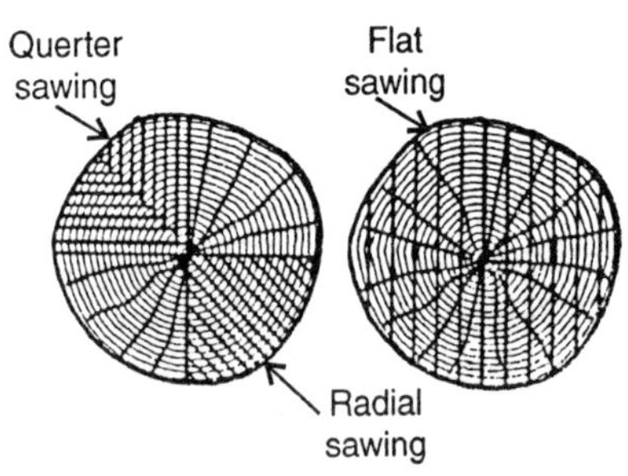

Fig. 5.2: Methods of cutting logs.

If we look the log of wood in a traverse section, it looks like a series of concentric circles due to annual rings. The tangential plane is a plane at a tangent to these circles; the radial plane follows a diameter and passes through the centre. Planks sawed tangentially to the annual rings are termed *flatsawn*. Giving a *flat grain*, while radially cut planks are termed *quarter- sawed*, giving an *edge grain*. If the annual rings are approximately at 45° to the face of the plank, the condition is called *angle grain. Cross grain* refers to a plank whose fibres are not parallel to the long axis of the plank. For seeing the methods of cutting logs, see Fig. 5.2.

NATURAL AND ARTIFICIAL SEASONING OF WOOD

By seasoning of wood, it makes the timber lighter in weight, more resilient, and less liable to twist, warp, and split. It is also in a better condition to retain its size and shape after being made into a piece of joinery. Wood increases in strength, hardness and stiffness as it dries. The methods of seasoning are as follows:

Natural seasoning: This is called an drying. In this method the balks (roughly squared logs) are stacked under cover with spacers in between, so that a free circulation of air is provided all round them. This method is slow, but gives the best results. A further period of seasoning should take place after the balks are sawn up and converted into planks or boards. This is to help dry out the interior of the timber which has been exposed by sawing.

Artificial seasoning: In this method the period of seasoning is very much reduced, a matter of two or three weeks being sufficient, according to the size or species of timber to be seasoned. The timber is stacked on a special truck and wheeled into a chamber which is then sealed. Hot air is circulated by fans, and a certain amount of steam is added in order to retain the correct humidity. Samples are tested at intervals to ascertain the percentage of moisture remaining in the timber. Seasoned timber still contains a proportion of moisture, which varies from 16 to 22 per cent according to the seasoning conditions, and this need not be dried out any further if intended for use out-of-doors.

In ease the timber is used for interior work or in a heated atmosphere, the timber should be further conditioned, that is, dried in warm-air kilns, or stored in a similar atmosphere to that in which it will be fixed, until the moisture contents is brought down to the region of 8-12 per cent.

Moisture content is the ratio of amount of water in a sample which shows or indicated as a percentage.

DIFFERENT TYPES OF DEFECTS IN TIMBER

Seasoning defects: As the moisture evaporates during

seasoning, shrinkage of the timber takes place. If a balk is dried too quickly, splits and cracks will appear. Shrinkage in the length is negligible, but it is more pronounced in the direction of the annual rings.

Between Sapwood and Heartwood, sapwood shrinks more than *heartwood*, so that a board cut from the outside of a log will shrink more, and have a greater tendency to warp, than one cut from the centre. The timber stored after felling the tree will also undergo the above shrinkages and may show *radial splits* in dry and hot environment.

If the timber is stared in a wet air, it can't be used or connected in a piece of joinery and then fitted into a warm room. The result will be rapid shrinkage, warping, twisting and splitting.

During seasoning, due to uneven drying, the timber may twist. Fig. 5.3(a) shows three types of defects: crook, bowing and end splits.

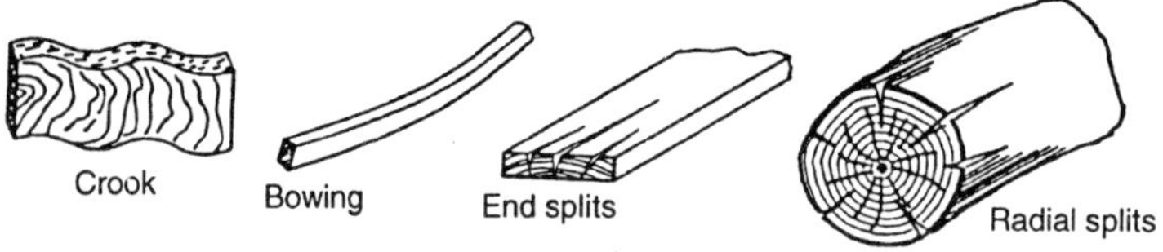

Fig. 5.3 (a) Seasoning defects.

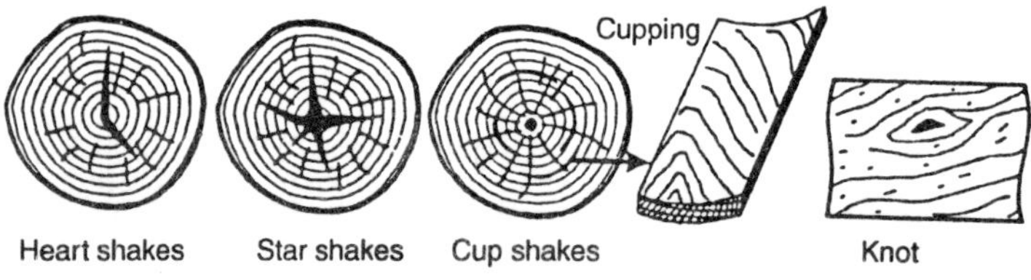

Fig. 5.3 (b) Natural defects.

Natural defects: In the standing tree, heart shakes are single splits and the same are invisible till the tree is felled. A number of such shakes radiating from pith is known as *star shakes*. If annual rings separate, cup shake is formed. The growth of branches also form a knot distorting the grain of the timber. You may see some of the natural defects in above figures.

Another defects in the timber when resinous materials accumulate with in the wood is Pitch.

Defects due to destructive agents include *worm holes* and *fungi-decay*. Wormholes are small holes in the wood caused by insects (termite, marine-borer and beetles) boring through the wood.

There are number of wood destroying *fungi* which attack standing timber, felled timber and conditioned timber.

Out of defects due to manufacturing/cutting, some of the manufacturing defects are (1) cross grain, (2) machine burn.

Wood in which the cells/fibres run at an angle with the axis, or sides, of the piece is defined as cross-grained.

Machine burn is usually caused by planer blades that are dull or spur and can be seen as dark streaks along the face of the boards.

ISO classification: As per classification by the International Organisation for Standardization-ISO the defects as per ISO 2299-1986 are as follows:

2.1 Knots, 2.2 Shakes, 2.3 Irregularities of wood structures and abnormal colourisation of wood, 2.4 Defects caused by fungi, 2.5 Defects caused by insects, 2.6 Sawing defects and deformation.

Each of the classes is further divided to subclasses and each of the subclasses is further divided in sub-subclasses.

WOOD'S CLASSIFICATION AND CONVERSION

Classification

Timbers, which are used for commercial purposes, are divided into two following classes:

(1) soft wood, and

(2) hard wood. These two terms, however, have no reference to the hardness of the wood and they are only two botanical classifications.

The woods which have long narrow leaves are observed as soft woods. They contain turpentine and resinous matters in their cells. The average soft-wood contains about 42 per cent cellulose, 25 per cent hemicellulose, 30 per cent lignin and 3 per cent miscellaneous items. Lignin also known as 'wood glue' holds the other items together in the wood. It can be converted into vanillin or other resinous materials useful for foundry mould. Soft woods are light in weight and light coloured, have distinct annual rings but not visible medullary rays, and the colour of the sapwood is not distinctive from their heartwood. The fibres are generally coarse but straight, and hence, capable of resisting direct axial stresses; but they cannot resist any kind of stress developed across their fibres and the timber gets splitted easily.

The woods obtained from broad-leaved trees are hard woods. It contains about 45 per cent cellulose, 25 per cent hemicellulose, 23 per cent lignin and 7 per cent miscellaneous items. The annual rings are more compact, thin and less distinct, but the medullarly rays are visible in most, and in some cases very pronounced. Hard woods are darker in colour, comparatively heavy. The fibres are fine grained, compact, properly bonded, and often found very straight. So hard woods are nearly equally strong both along and across the fibres and can resist axial stress as well as transverse strain, shock and vibration quite satisfactorily.

Some of hard woods which do not catch fire like Sal, Pyingads and Ash are classified as refractory. On the other hand, some of the soft woods which catch fire easily like Pine, Deodar and Fir are classified as non–refractory because of the presence of resinous matter.

Cutting of Log or Conversion

The cutting of the log into usable pieces of timber is called conversion. The timber is available in market in following forms:-

Log	-	Felled tree
Balk	-	Log squaring up
Planks	-	(275 to 450) mm × (75 to 150)mm
Deals	-	225 mm × 50 to 100 mm
Batten	-	135 mm × 150 mm
Quartering	-	2 × 25 mm^2 unto 150 × 150 mm^2 stuff
scantling	-	75 mm × 75 mm, 100 mm × 50 mm, 100 mm × 75 mm etc.

VARIETIES OF TIMBER IN INDIA

There are several varieties of Indian timbers as:

Mahogany. The wood is of red brown colour, very durable when kept dry. Usually, it has fine, wavy grains, and uniform colour. It contains resinous oil which prevents attack of insects. They are available in Himalayas and used for pattern-making and cabinet work.

Babul. The wood is pale red to brown in colour, close-grained, hard and tough, but elastic, and takes a good polish. They grow abundantly all over India and are used for bodies of carts and wheels, agricultural implements, tool-handles, etc.

Mango. The wood is of inferior quality, coarse and open grained and of deep gray colour. They decay readily when exposed to moisture and are greedily eaten by white ants. They are largely found all over India, and being plentiful and cheap, widely used for common doors, windows, and furniture.

Teak. The wood is brown in colour, straight-grained, and is fragrant when freshly cut, very strong and durable, yet light and easily worked. It shrinks little, takes a smooth polish, and can be seasoned quickly. They are available in large quantities in Burma, Malabar and Central India, and suitable for practically every description of work.

Sissu. The wood is dark brown in colour, tough, durable and has well-marked coarse grains. It is one of the best Indian woods for joiner's work—tables, chairs, and other furniture and is widely distribute in Northern and Northern and Central India.

Sal. The wood is of a dark brown colour, hard, close-grained heavy, resistant to white ants and durable. It seasons slowly, is hard to work and does not take a high polish. They grow abundantly in the forests at the foot of the Himalayas—in UP, Bihar, and Assam—also in Central India and South India and largely used for constructional purposes.

DIFFERENT CARPENTRY TOOLS

A large number of tools are used during working to accurate shapes and dimensions, the wood-worker must know the use of a large number of tools. The main types which are manipulated by hand are described and illustrated as under:

1. Measuring tools.
2. Cutting tools.
3. Planing tools.
4. Boring tools.
5. Striking tools.
6. Holding & miscellaneous tools.

MEASURING TOOLS

There are several measuring tools by which accurate work may be assured. The followings are such types of tools:

For measuring and setting out dimensions, rules of various sizes and designs are used by carpenter but they usually work with a *four-fold box-wood rule* ranging from 0 to 60 cm. This is graduated on both side in millimetres and centimetres, and each fold is 15 cm long. All the four pieces are joined with each other by means of hinged joints which make the scale folding.

A long and flexible tape is used to measure the larger measurement. Such rules are very useful for measuring curved and angular surfaces. When not in use, the blade is coiled into a small, compact, watch-size, case.

Straight edge. The straight edge (Fig. 5.4) is a machined flat piece made of wood or metal having truly straight and parallel edges. One of the longitudinal edges is generally made levelled. This is used to test the trueness of large surfaces and edges.

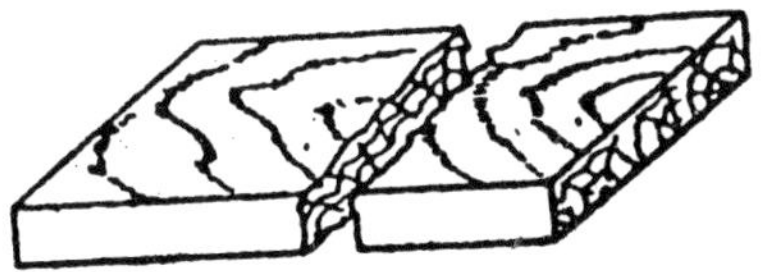

Fig. 5.4: Straight edge.

Try square. Try squares (Fig. 5.5) are used for marking and testing angles of 90°. It consists of a steel blade, riveted into a hard wood stock which has a protective brass plate on the working surface. Another type is the all-metal square, with steel blade and cast iron stock. Sizes vary from 150 to 300 mm, according to the length of the blade.

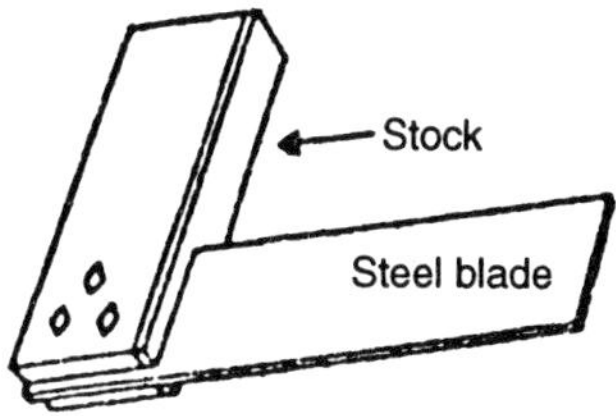

Fig. 5.5: Try Square.

Mitre square. These are used to measure an angle of 45°. (Fig. 5.5.) They are made of all metal with a nickel-plated finish or with a steel blade, and an ebony or rose-wood stock. The blade varies from 200 mm, 250 mm and so on to a maximum of 300 mm long.

Bevel square. It is similar to the try square but has a blade that may be swivelled to any angle from 0 to 180° as shown in Fig. 5.7. This tool is adjusted by releasing with a turn screw of suitable size in a machine screw running in a slot in the blade.

Combination square. Combination square is a combination of square 45 degree bevel, set square, rule, straight edge and centre finder. Some wood workers prefer a combination square which is similar to the combination set used in bench work.

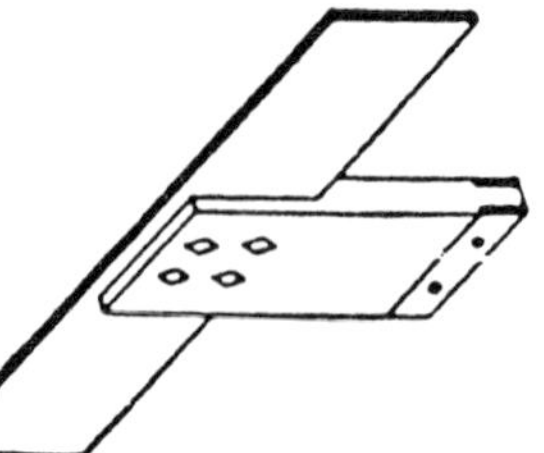
Fig. 5.6: Mitre square

Marking knife. These knives (Fig. 5.8) are used for converting the pencil lines into cut lines. They are made of steel having one end pointed and the other end formed into a sharp cutting edge.

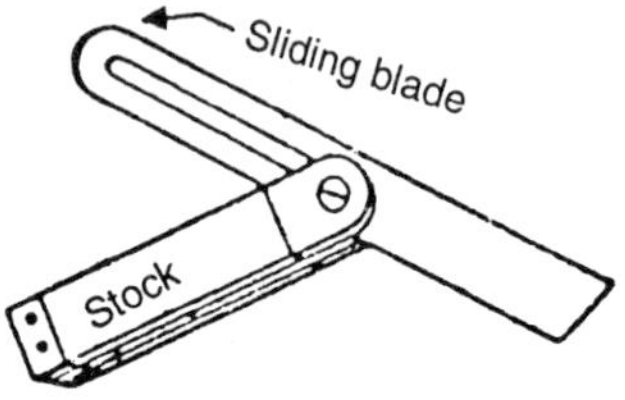

Fig. 5.7: Bevel square

Gauges. Gauges are used to mark lines parallel to the edge of a piece of wood. It consists of a small stem sliding in stock. The stem carries one or more steel marking points or a cutting knife. The stock is set to the desired distance from the steel point and fixed by the thumbscrew. The gauge is then held firmly against the edge of the wood and pushed along the sharp steel point marking the line. There are three type of gauges:

Fig. 5.8: Marking knife

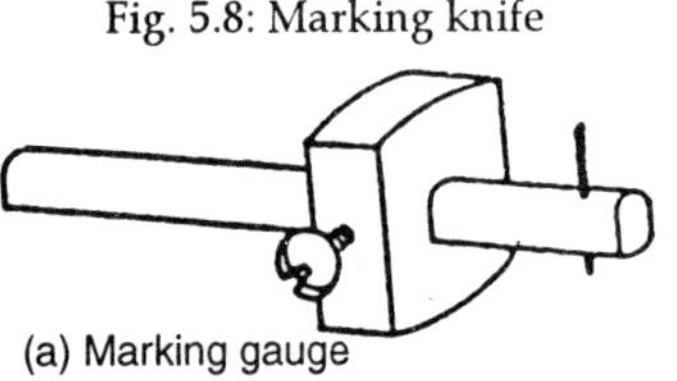
(a) Marking gauge

(1) The *marking gauge* (Fig. 5.9a) has one marking point. It gives an accurate cut line parallel to a true edge, usually with the grain. The *panel gauge*, is longer than the marking gauge, and is used to gauge lines across wider surfaces.

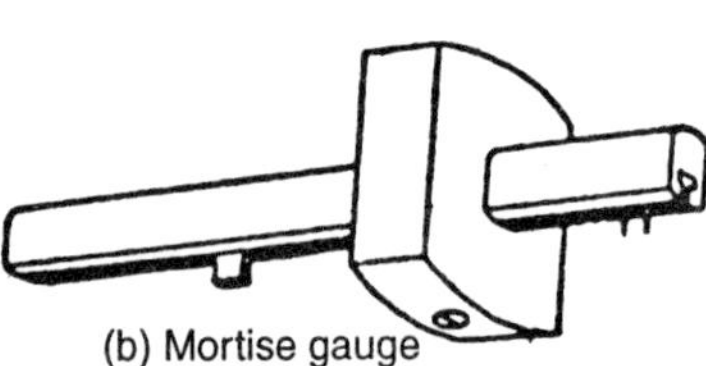
(b) Mortise gauge

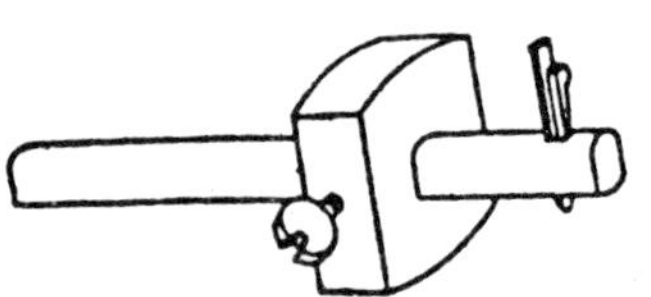
(c) Cutting gauge

Fig. 5.9: Ganges

(2) The *mortise gauge* (Fig. 5.9b) has two marking

points-one fixed near to the end of the stem and the other attached to a brass sliding bar. These two teeth cut two parallel lines, called mortise lines.

(3) The *cutting gauge* (Fig. 5.9c) has a cutting knife held in position by a wedge so that its projection may be varied for the depth of the cut. This gauge is useful for gauging fine deep lines for such joints as dovetails on wide wood, for cutting the edges of grooves, for inlaying, cutting through very thin stuff to make small strips, and cutting small rebates.

Wing compass. Wing compasses are composed of two finely pointed steel legs which are set to the desired position and held by a set screw and quadrant. They are used when stepping off a number of equal spaces, marking circles or arcs, and when scribing parallel lines to straight or curved work.

Trammel: It is a form of beam compass, with a wooden beam, to take in work that is beyond the scope of a compass.

Divider. Dividers have both points sharpened in needlepoint fashion for dividing out centres.

Caliper. Calipers are used for measuring outside and inside diameters etc., especially where the sectional measurements cannot be taken.

Spirit level and plumb bob. By this tool, the position of large surface is tested. The plumb bob tests for vertical position. A combination of these two gives a right angle, and they are used where a try square would be far too small.

CUTTING TOOLS

There are several cutting tools like saws, chisels, and gouges.

Saws. It is very common and most useful tool which is used to cut across the grain. When cutting across the grain, a different action is required from the saw teeth than when ripping with the grain. Therefore, different types of saws are used, as one type cannot do both jobs successfully. A saw is generally specified by the length of its blade measured along the toothed edge, and pitch of teeth, expressed in millimeters. Fig. 5.10 shows the different types of saws in common use.

Rip Saw: By this tool the grain is cut along in thick wood. The blade is made of high grade tool steel, and may be either straight or skew backed. It is fitted in a wooden handle made of hard wood by means of rivets or screws. Rip saws are about 700 mm long with 3 to 5 points or teeth per 25 mm.

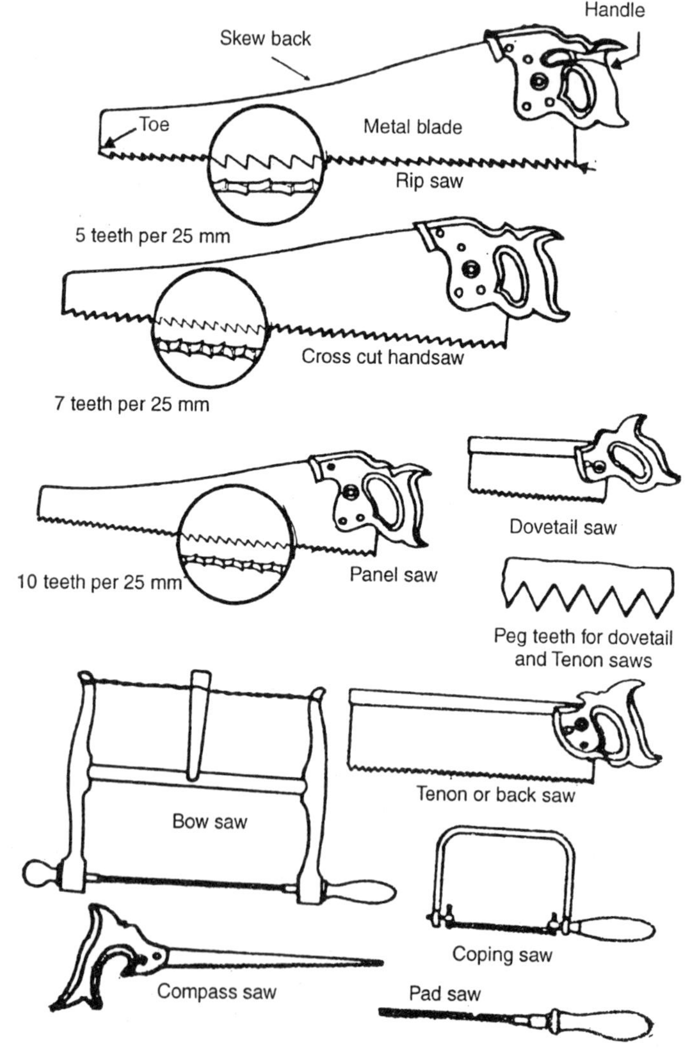

Fig. 5.10: Different types of saws.

The leading edge of the teeth forms a right angle with a line joining the points, and should be filed squared across the saw, with no bevel on front or back of the tooth. The action of these teeth is that of a series of chisels, which tear out shavings each equal to the width of a tooth.

The teeth are bent alternately, one to the right, the next to the left. Bending the teeth in this manner is called "setting in saw". The set of a saw provides clearance to prevent the blade from binding during the sawing operation.

Cross-cut saw. Cross cut saws are used for cutting *across* the grain in thick wood. They are 600 to 650 mm long with 8 to 10 teeth per 25 mm. The action of the teeth is that of a series of knives which is the fibres and force out the waste wood in the form of saw dust.

Panel saw. This is about 500 mm long with 10 to 12 teeth per 25 mm and is very much like the cross-cut saw. It has a finer blade and is used for fine work, mostly on the bench. This is often used for ripping as well as cross-cutting. The teeth have slightly more hook than those of a cross-cut saw.

Back saw. Back saw is mostly used for cross-cutting when a finer and more accurate finish is required. The blade, being very thin, is reinforced with a rigid steel back. Tenon saw blades are from 250 to 400 mm in length and generally have 13 teeth per 25 mm. The teeth are shaped in the form of an equilateral triangle and are sometimes termed "peg" teeth.

Dovetail saw. Where the great accuracy is needed, dovetail saw is used which is a smaller version of the tenon and fine shallow cuts are to be made. The number of teeth may be from 12 to 18 per 25 mm, while the length may vary from 200 to 350 mm.

Bow saw: It has 250 to 350 mm long blade in a wooden frame. The blade is held in tension by twisting the string with a small wooden lever. These saws are used for cutting quick curves, and, as the handles revolve in their sockets, the blade can be adjusted to any desired position when in use.

Coping saw. It has a very similar blade, held rigid in spring-metal frame. The blade is tensioned by screwing the handle. This saw is used for small radius curves.

This is used for sawing small curves in confirmed spaces and has a narrow tapering blade about 250 to 400 mm long, fixed to an open-type wooden handle. There are two types of compass saw, one having a fixed blade and the other with three interchangeable blades of different widths.

Keyhole saw: Keyhole saw is the joiner's smallest saw. The blade being about 250 mm long. The blade of the pad saw is secured to the handle, through which it passes, by two screws. This arrangement allows the blade to be adjusted to the best length required according to the work. This saw is used for cutting key holes, or the starting of any interior cuts.

Chisels. Wood chisels most commonly in use include firmer chisels, either square of bevel edged, paring chisels, and mortise chisels.

Specification by way of width and length of the blade is as follows:

Firmer chisel. This is most useful for general purpose as shown in Figure 5.11. and may be used by hand pressure or mallet. It has a flat blade about 125 mm long. The width of the blade varies from 1.5-50 mm.

Beveled edge firmer chisel. This is used for more delicate or fine work (Fig. 5.12). They are useful for getting into corners where the ordinary firmer chisel would be clumsy.

Paring chisel. Both firmer and beveled edge chisels when they are made with long thin blades are known as paring chisel (Fig. 5.13). This is used for shaping and preparing the surfaces of wood and is manipulated by the hands. The length ranges from 225 to 500 mm and width from 5-50 mm.

Mortise chisel. The mortise chisel shown in Fig. 5.14 as its name indicates, is used for chopping out mortises. These

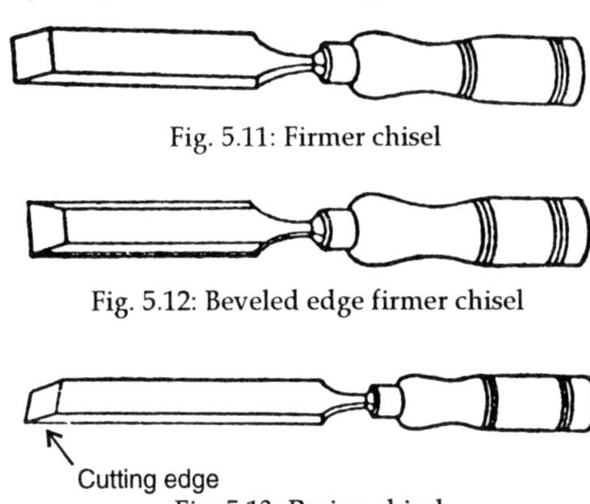

Fig. 5.11: Firmer chisel

Fig. 5.12: Beveled edge firmer chisel

Fig. 5.13: Paring chisel

Fig. 5.14: Mortise chisel

chisels are designed to withstand heavy work. They are made with a heavy deep (back to front) blade with a generous shoulder or collar to withstand the force of the mallet blows on the oval-sectioned handle. Many mortise chisels are fitted with a leather washer at the shoulder to absorb the hard shocks of the mallet blows. Blades vary in width from 3-16 mm.

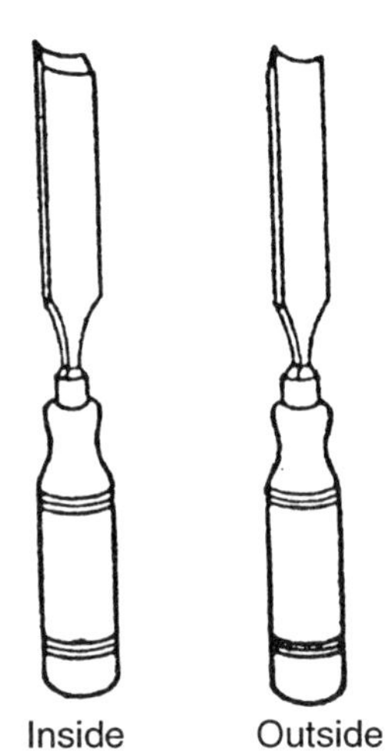

Fig. 5.15: Gouges

Gouges: It may be inside or outside ground which are the chisels with curved sections. Inside ground gouges are used in exactly the same way for inside curved edges as a chisel would be for straight one; outside ground gouges are used for curving hollows. Outside ground gouges are known as *firmer gouges* and inside ground gouges are called *scribing gouges*. When the later are made long and thin they are *paring gouges*. Gouges are made to a large number of different curves for different work, and the size ranges from 6 mm, with intermediate sizes to a maximum of 40 mm wide.

PLANES

These planes can be likened to the chisel fastened into a block of metal or wood, and its blade cuts exactly like a wide chisel. The planes, in general use, are the jack, trying, and smoothing planes, and are known as bench planes. Besides, there are other planes which are used for special work.

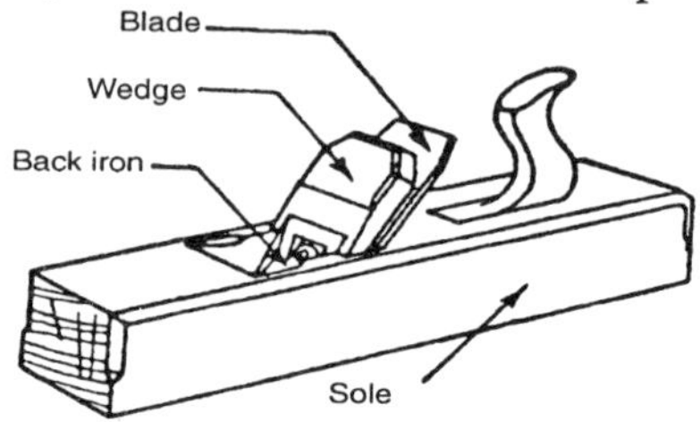

Fig. 5.16: Jack plane

Jack plane. This is the commonest and is used for the first truing-up of a piece of wood as showing in Fig. 5.16.

Trying plane. It is called a finishing plane, and is set with a very fine cut. It consists of a block of wood into which the blade is fixed by a wooden wedge. The blade is set at an angle of 45° to the sole. On the cutting blade another blade is fixed called cap iron or back iron. This does not cut, but stiffens the blade near its cutting edge to prevent chattering and partially breaks the shaving as it is made. It is the back iron which causes the shavings to be curled when they come out of the plane. Some types of planes do not have a cap iron. Jack planes are obtainable from 350 to 425 mm in length and with blades 50 to 75 mm wide.

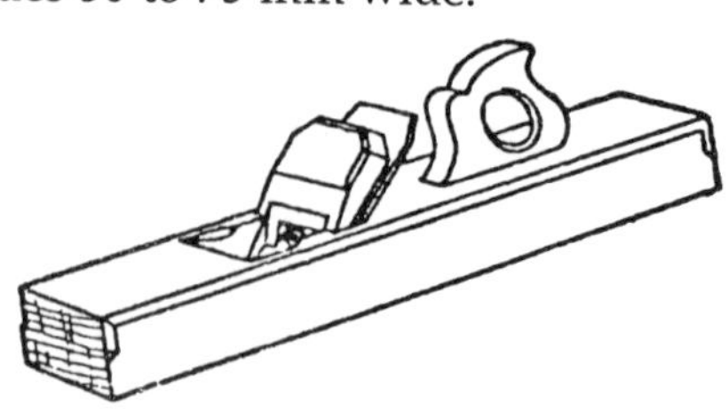

Fig. 5.17: Trying plane

It is used for producing as true surface or edge as possible, and is set to cut a shaving as thin as the smoothing plane. The length of the plane varies from 550 to 650 mm and the section of the body is 85 mm by 85 mm, with irons 60 mm wide.

Smoothing plane. This is similar in action to a jack plane but it is set to cut a much thinner shaving. A smoothing plane, as its name indicates, is used for smoothing or finishing after a jack plane. The cutting edge of the latter is slightly curved, but a smoothing plane has a straight cutting edge. It is 200 to 250 mm long having a blade of 70 mm wide.

1. **Rebate plane.** A rebots forms a ledge which is used for positioning glass in frames and doors, this forms a ledge which is used for positioning glass in frames and doors. The rebate plane shown in Fig. 5.19 is used for sinking one surface below another, and shouldering one piece into another. The blade is open at both sides of the plane, and must be perfectly straight at the cutting edge. Widths range from 12 to 50 mm.

2. **Plough plane.** Plough is used to fit it into a groove, not into a rebate, as and when a panel is needed in a door. The plough plane illustrated in Fig. 5.20 is used to cut these grooves. The depth of groove is controlled by a depth gauge which is fixed on the body of the plane and operated by a thumbscrew. These planes are usually supplied with eight

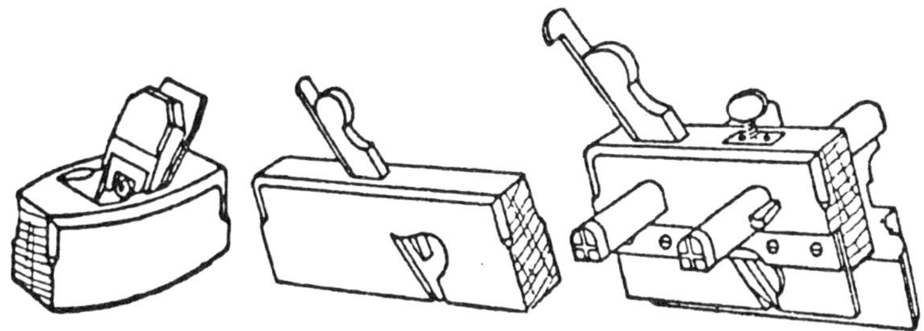

Fig. 5.18: Smoothing plane. Fig. 5.19: Rebate plane. Fig. 5.20: Plough plane.

to nine blades, vary in width from 3 to 15 mm and, of course, they are all interchangeable.

3. **Spokeshave.** Spokeshave is used for cleaning up quick curves (as per Fig. 5.21). The same is a form of small

plane. There are two types, one which has a flat sole for outside curves and one which has a curved sole for inside curves. Now-a-days, spokeshaves are made of iron, and some have a screw adjustment for the amount of cut.

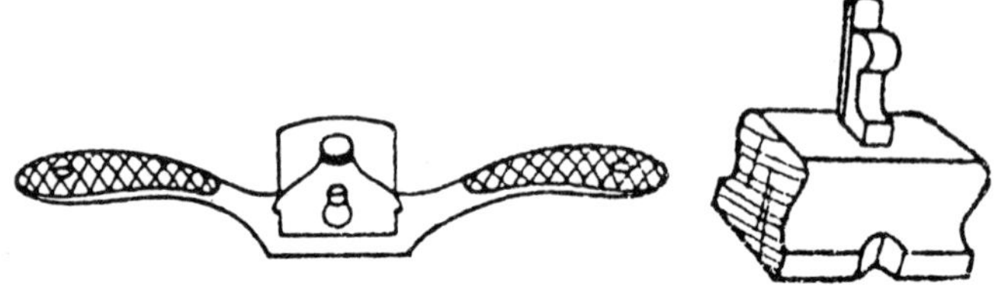

Fig.5.21: Spokeshave. Fig. 5.22: Router.

4. Router. The Router is used for cleaning out and levelling the bottom of grooves or trenches to 9 constant depth after the bulk of the waste material has been taken out with saw and chisel.

5. **Metal plane.** It is used for the same purpose like wood plane but facilitate a smoother operation and better finish. The body of a metal plane is made from a gray iron casting, with the side and sole machined and ground to a bright finish. The thickness of the shaving removed is governed by a fine screw adjustment, and a lever is used for adjusting the blade at right angles. A metal jack plane us shown in Fig. 5.23.

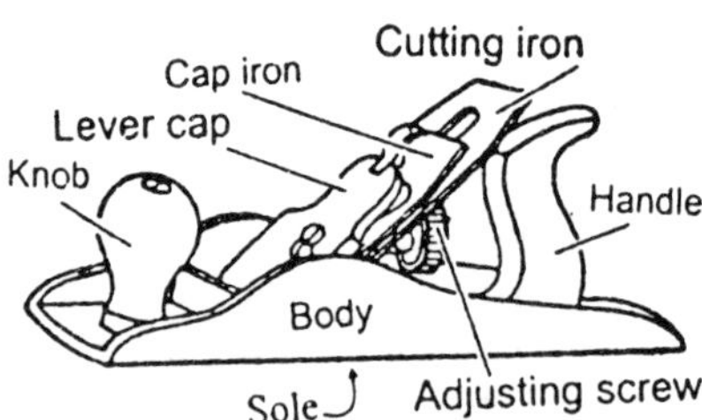

Fig. 5.23: Metal jack plane.

6. **Special plane.** There are a number of special planes other than the above-mentioned. These are used by the woodworker to do special work. They include *compass* or *circular plane* for planing curves; *bull nose rebate plane* for cleaning into rebates and corners inaccessible with other

planes; *shoulder plane* for planing across the end grain or handwood shoulders; *block plane* for planing small parts, especially when model making; and *moulding plane* for producing a particular size and shape of moulding.

BORING TOOLS

These are frequently necessary to make round holes in wood, and they are selected according to the type and purpose of the hole. They include bradawl, gimlet, brace, bit and drill.

Bradawl and gimlet. These are hand-operated tools as illustrated & in Fig. 5.24 and are used to bore small holes, such as for starting a screw or large nail.

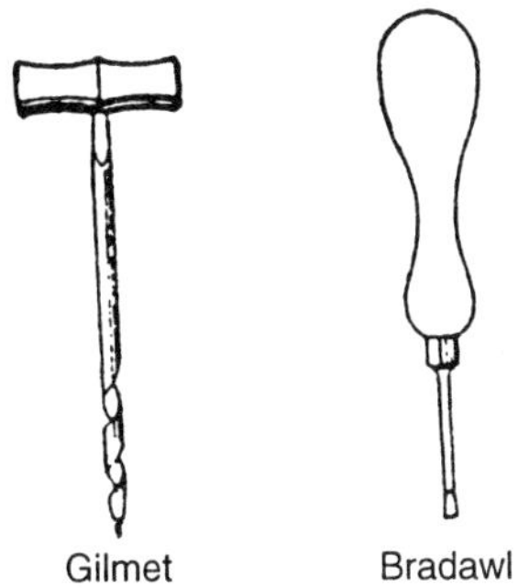

Fig. 5.24: Bradawl & gimlet.

Brace. This tool is used for holding and turning a bit for boring holes. It has two jaws, which grip the specially shaped end of the bit. There are two types of braces in common use– ratchet brace and wheel brace. The *ratchet brace* is most useful for turning bits and drills of all kinds, being adaptable (a) for working in confined situations, and (b) for when the cut is particularly heavy and it is desirable to pull the handle through a quarter-turn only. A ratchet brace is shown in Fig. 5.25.

The *wheel brace* (Fig. 5.26) is used to hold round and parallel-shanked drills. This tool is invaluable for cutting small hole, accurately and quickly.

Bit. Bit is a main organ of boring tools. The common types of bits used are shown in Fig. 5.27 and described below:

Shell bit. Shell bit is used for boring holes unto 12 mm diameter and which do not require a high degree of finish or size.

Twist bit. Twist bit has a twisted stem and screw point. This bit produces a long, clean, and accurate hole either with or across the grain. This may be obtained in sizes from 6 to 35 mm diameter. The shorter type is called 'dowel' bits and is used for preparing true and accurate holes to receive dowels.

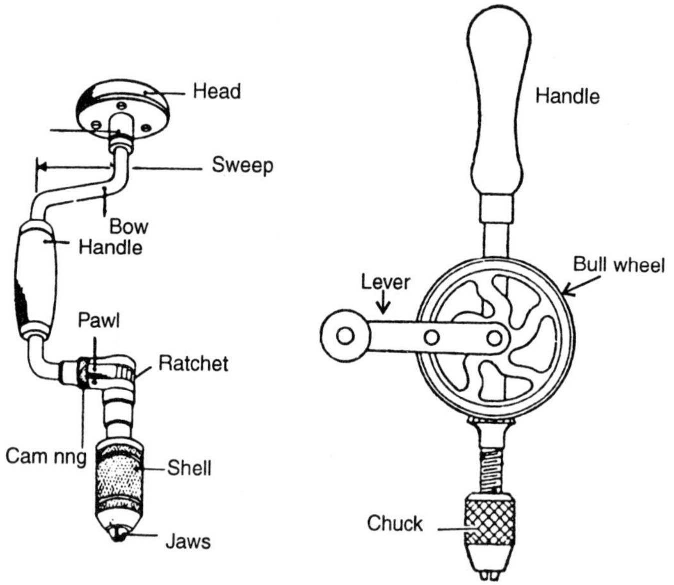

Fig. 5.25: Ratchet brace

Fig. 5.26 Wheel brace

Expeansive bit: In this bit the main cutter can be adjusted to varying diameter within a certain range. It is fixed to the desired mark on the scale, and clamped in position by the plate and screw. Expansive bits are made in four sizes with interchangeable cutter for boring holes from 12 to a maximum of 125 mm diameter.

Centre bit. This is one of the common bit which is used for forming shallow holes across the grain. Centre bits produce an accurate and clean hole and may vary from 3 to 35 mm in diameter.

Forstner bit. Forstner bit is extremely useful for sinking clean hole partly through a piece of wood and for cleaning out recesses. It has a small centre point for commencing and is then guided by its outer rim.

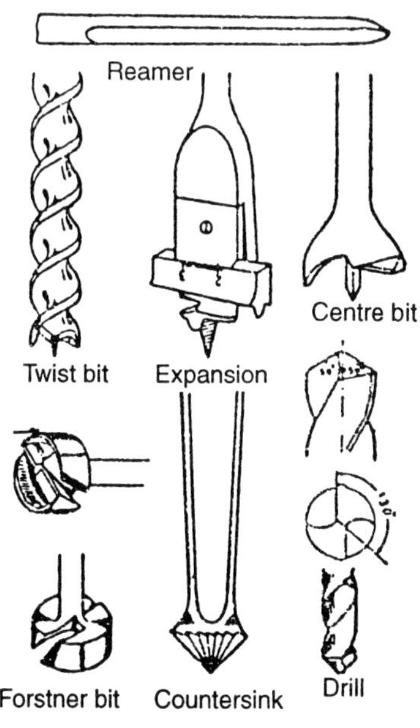

Fig. 5.27: Types of bits.

Countersink bit. This is used to shape a hole to fit the head of a countersunk headed screw. Other then the foregoing there are, of less importance, *nose bit, spoon bit, lip* and *spur bit, screw driver bit,* etc.

Drill. To make screwholes, drills are very convenient especially when used with a wheel brace. This is adapted for drilling holes when wood working bits would be spoiled. *Reamers* are tapered bits shaped like shell bits and used for enlarging holes.

STRIKING TOOLS

Striking tools include hammers and mallets.

Hammer: There are many type of hammers. The engineers use ball-peened hammers, woodworks crosspeened and claw hammers. The *Warrington hammer* shown in Fig. 5.28 is the type mostly used for bench work and all light jobs. The head is cast steel, the face and peen being tempered; the shaft which is wedged tightly into the head is made of wood or bamboo. These hammers are identified by size numbers and weight, No. 00, 200 gm up to No. 6,550 gm. The carpenter more often favours the *claw hammer* (Fig. 5.29) because it serves the dual purpose of a hammer and a pair of pincers. The claw is used for pulling out any nails accidentally bent in driving. These hammers are made in numbers sizes from 1 to 4, weighing 375, 450, 550 and 675 gm.

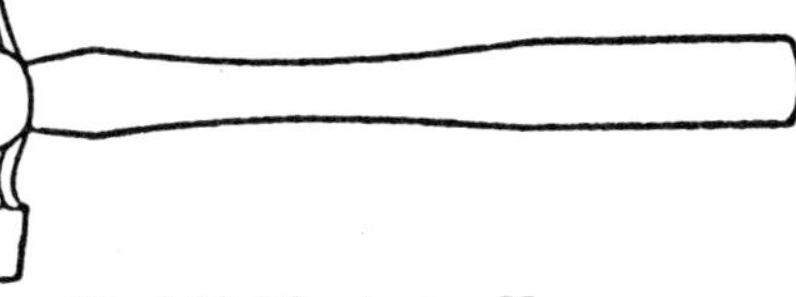

Fig. 5.28: Warrington Hammer

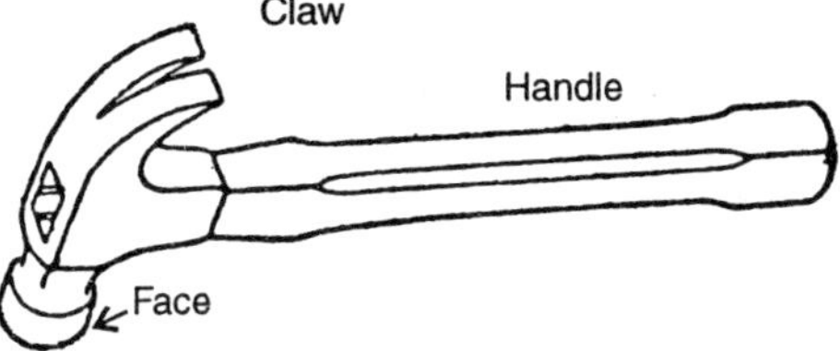

Fig. 5.29: Claw Hammer

Mallet Hammer. This is a wooden-headed hammer of round or rectangular shaped (see Fig. 5.30). The striking face is made flat to the work. A mallet is used to give light blows to the cutting tools having wooden handle such as chisels, and gouges.

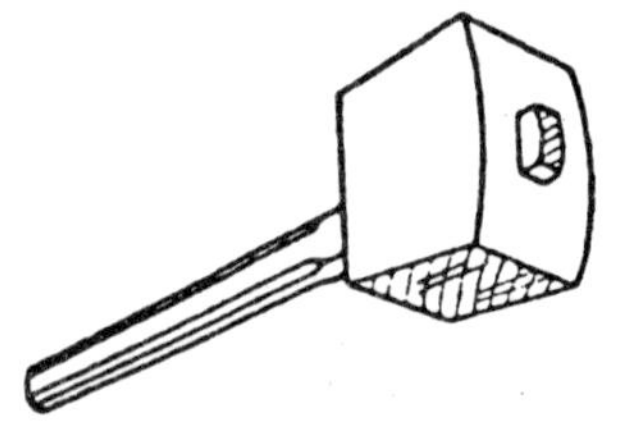

Fig. 5.30: Mallet

HOLDING TOOLS

To cut wood accurately, the wood workers used the holding tools or devices. Some of them are given below:-

Bench vice: As shown in Fig. 5.31, the Bench vice is very useful to wood workers. Its one jaw is fixed to the side of the table while the other is kept movable by means of a screw and a handle. The whole vice is made of iron and steel, the jaws being lined with hardwood face which do not mark and which can be renewed as required.

Bench stop. It is a block of wood projecting above the top surface of the bench. This is used to prevent the wood from moving forward when being planed.

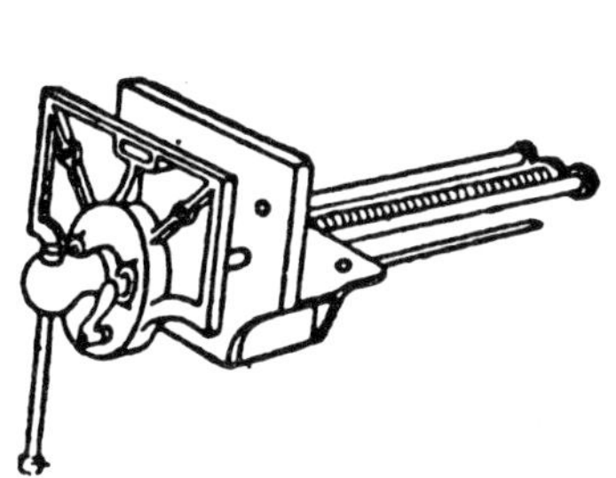

Fig. 5.31: Bench vice

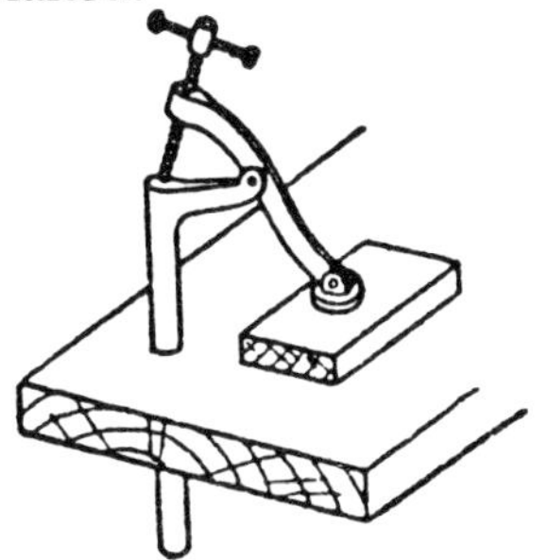

Fig. 5.32: Bench holdfast

Bench holdfast. This is made of cast iron pillar, square-cut screw threads on a steel bar, as shown in Fig. 5.32, with a light vice handle and a drop-forged steel arm. By boring a series of holes through the top of the bench, holdfasts can secure the work in any desired position. This is useful for holding a piece of wood down on the bench when a vice is not advisable.

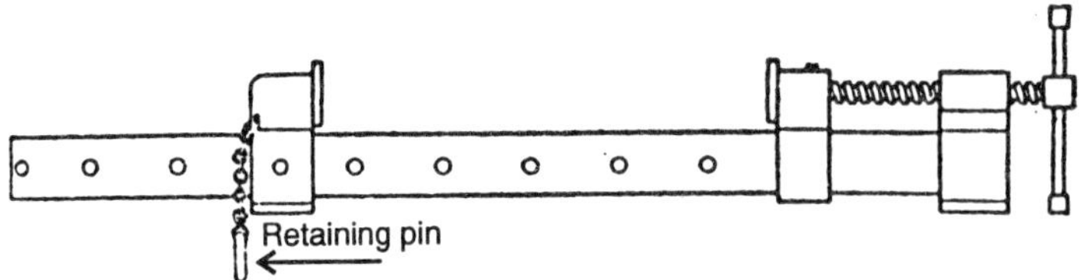

Fig. 5.33: Sash cramp

Sash cramp. It is also called bar cramp which is made of a steel bar of rectangular section, as shown in Fig. 5.33., with malleable iron fittings and a steel screw. This is used for holding wide work such as frames or tops.

G-cramp. G-camp is used for smaller work (see Fig. 5.34). It consists of a malleable iron frame that can be swivelled and a steel screw to which is fitted a thumbscrew.

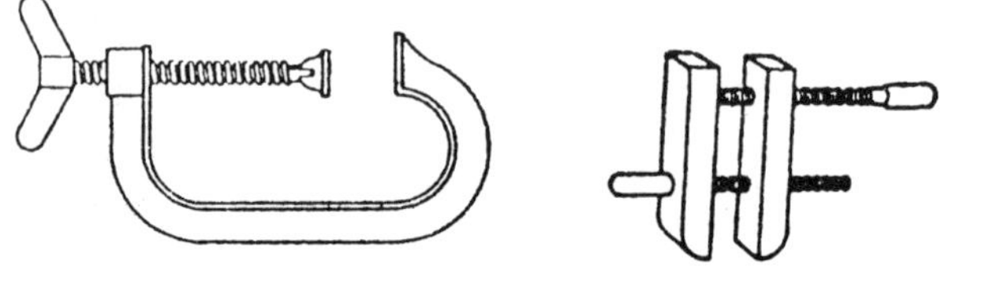

Fig. 5.34: G-cramp

Fig. 5.35: Hand screw

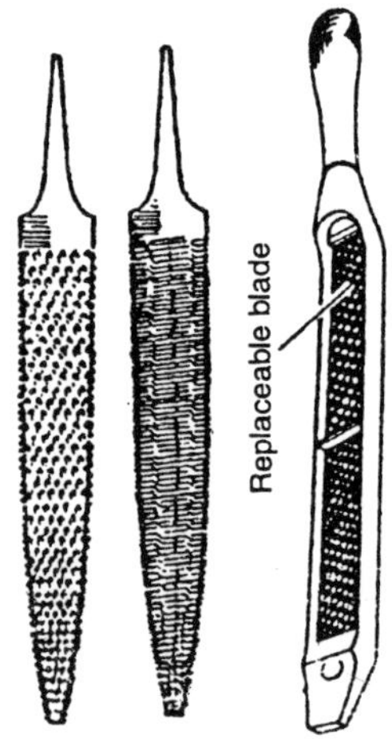

Rasp File Surform tool

Fig. 5.Rasp, File and Surform tool

Hand screw. Where a wider area of pressure than that provided by a G-cramp is require, the hand screw in Fig. 5.35 is used. It consists of two steel screws fitted to two jaws made of wood.

OTHER TOOLS

There are some other tools like Raps and Files, Scraper, oilstone, glass-paper, pinear, screw driven, etc. Their uses are mentioned below:

Rasps and files. By these tools some curved surfaces are cleaned up. For instance, certain concave shapes are so small that the spokeshave cannot enter them and here a file is invaluable. Scratches left by the file can be removed with the scraper and glass paper. Surform tools introduced by Stanley Tools contain many small cutting teeth, each of which acts rather like a chisel or small plane. The teeth are not inclined to choke easily. The blades are disposable and when blunt must be replaced. Flat, convex and round blades are available. Illustrations are given in Fig. 5.36.

Scraper: Scrapers in Fig. 5.37 scrapes very small shavings off the wood. The scraper in Fig. 5.37 consists of a piece of thin steel, hardened and tempered. A fine the edge is made by pushing over or burnishing the edge of the metal to form that what is called a "burr"

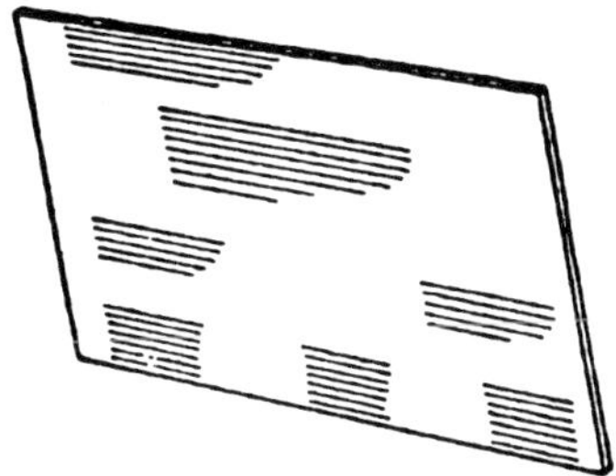

Fig. Scraper

Glass paper. Glass paper consists simply of small particle of glass stuck to sheets of paper. Where a surface is covered with innumerable minute imperfections so small that no other cutting tool will do, then glass-paper should be used. It is the sharp edges of these particles which cut the wood. Glass-paper is made in varying grades according to the size of glass particles. These grades are denoted by numbers such as 00,1, etc.

Oilstone: This is an essential part of a carpenter's kit of tools. Oilstones may be either natural or artificial. The best-known varieties of stones are carborundum, Washita, Turkey and India. These may be obtained in various grades known as coarse, medium, or fine.

Pincer: This is mainly used for pulling out nails, tacks, etc. It consists of two arms one arm has a ball end and the other arm has a claw end for levering out small tacks.

Fig. 5.38: Pincer

Screw driver: This is a very comman tool which is used for screwing or unscrewing screws used in woodwork. These may be obtained in various shapes and sizes but the one shown in Fig. 5.39, known as a "cabinet screw driver", is considered to be the best type.

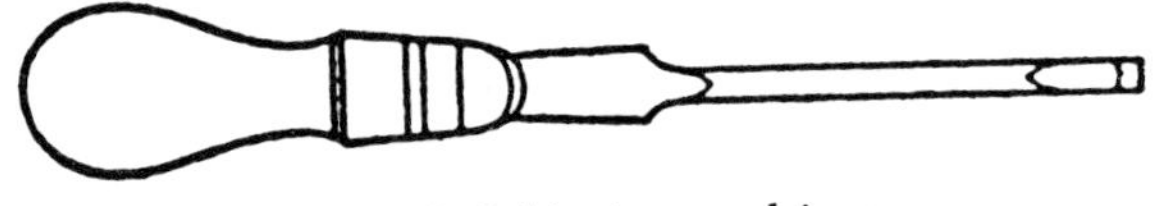

Fig.5.39: Cabinet screw driver

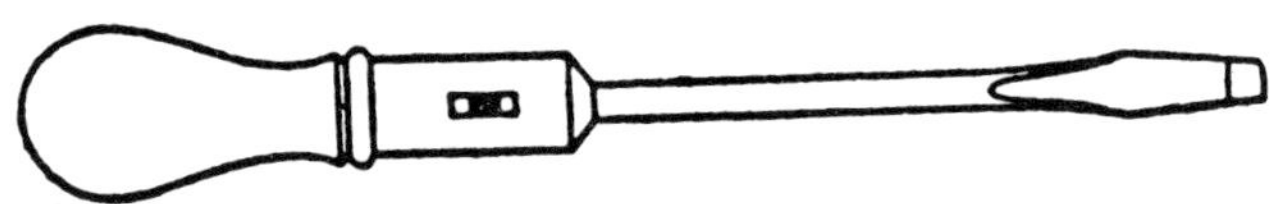

Fig. 5.40: Ratchet screw driver

HOW TO SET AND SHARP OF TOOLS?

Saws. These are reset by slightly bending each tooth the first tooth are way, the next the opposite way and so on. This is done so as to make the cut wider than the thickness of the blade, and ensure that it clears the sides of the cut as it passes through. A tool called *saw set* is used for this. There are different methods of setting the teeth of a saw but one by using a *gate* or *lever type hand set* is very common. It is shown in Fig. 5.41 and consists of a steel plate and handle. Along the edge of the plate are gaps or gates numbered according to the gauge or thickness of various saws. In use a suitable gap is slipped over the tip of a tooth, which is then bent over by using the handle of the set as lever.

The amount of set depends upon the type of work being

done. Green or damp timber will require that the teeth have plenty of set, whilst dry wood requires less. Normally, saws are sharpened after setting. A double-ended, single cut, three-cornered file, of suitable section size is the usual type of file used for sharpening saws. Great care must be taken to maintain the angle of the teeth as well a any cutting angle on the leading edge of each tooth.

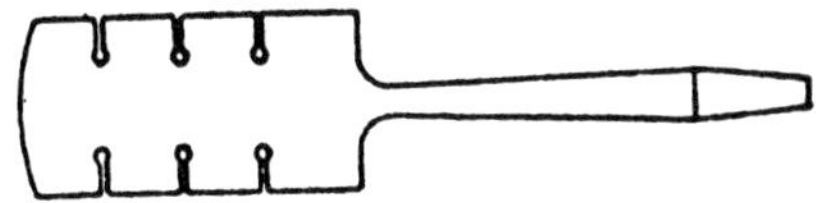

Fig. 5.41: Saw set (gate type)

The saw is held with the tooth projecting about 3 mm. Before commencing to file the teeth it is a good plan to pass a flat file lightly along the tops of the teeth to put the points in a straight line. The teeth can then be filed individually.

Planes. The efficient working of planes depends on several factors as follows:

1. The space between the cutting edge and the body of the plane has an important bearing on its working qualities. Coarse working planes, e.g., jack planes need at least 3 mm gap so that a thick shaving can bend easily through the opening. This gap may be as small as 0.4 mm for very thin shaving.
2. The angle at which the blade is set in the body of the plane fixes the *cutting angle*. In the case of planes with a back iron these are 47.5° and 42.5° respectively.
3. The distance between the cutting edge of the blade and the sole of the planes, called the *cut*, is kept from 0.8 mm to 1.6 mm for heavy cut and below 0.8 mm for fine cut. Of course, this depends on the quality of wood to be cut.
4. The back iron is set 3 mm to 0.4 mm from the cutting edge, depending upon the coarseness or fineness of the work. The nearer the back iron can be set to the cutting edge the better.

A grindstone and an oilstone are used to obtain fine cutting edges of planes.

The bevel is first obtained by being ground on a rotating grindstone at an angle of 25°. The tool is pressed lightly against the wheel which revolves towards the tool and the latter is moved sideways. Wet or dry grinding both may be employed during the operation. The edge obtained in grinding is very uneven because the particles of grit which do the cutting are larger in a grindstone.

The edge is, therefore, brought to a fine condition by being rubbed on an oilstone using oil. But instead of rubbing the whole of the ground bevel, the chisel is raised slightly so as to give a slightly larger angle of about 30 to 35°.

The smaller angle of 25° is known as the "grinding angle" and the larger one as the "sharpening angle". They are shown in Fig. 5.42.

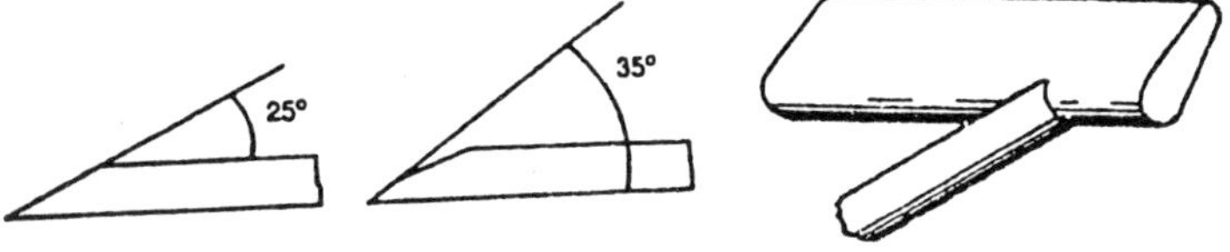

Fig. 5.42: Approximate grinding and sharpening angle

Fig. 5.43: Oilstone slip

For the sharpening of gouge, shaped stones (Fig. 5.43) called oilstone slips are used.

CARPENTRY PROCESSES

A number of hand operations to finish desired wood work is concerned with carpentry or joinery work.

The following is the principal processes used in wooden construction:

1. Marking.
2. Sawing.
3. Planing.
4. Chiselling.
5. Boring.
6. Grooving.
7. Rebating.
8. Moulding.

Indications or Marking

This is the process of setting out dimensions on a piece of wood for producing the required shape. These dimensions can be measured from an existing model or can be set out from the drawing prepared for the purpose. Each dimension is taken out with a folding rule which is the most convenient for general use, and is set out with the help of various instruments such as caliper, try-square, marking knife, marking gauge, etc. When marking a size the pieces are first planed true and square and then marked according to the desired dimension. The trueness of the surface is tested every time with the straight edge or the blade of the try-square or the steel rule. A zig zag pencil mark is made on the true surface to distinguish it from the other faces.

Sawing

This basic operation carried out in a carpentry shop. A wood is required to be sawn along the grains or across the grains and in many shapes such as straight inclined or curved. To start the cut, the thumb of the left hand is placed against the blade. This steadies the blade, enabling it to start in the right place, and prevents any accident in the event of the saw jumping. One or two short movements are first given, taking care that the saw works in the right direction and then full, easy strokes are applied to cut the wood. A point to note in all sawing is that the cut is made on one side of the line already marked and that is on the waste side, as in Fig. 5.44. A saw should never be forced and it is kept moving steadily for nearly its full length. Its own weight plus the slightest pressure is all that is needed.

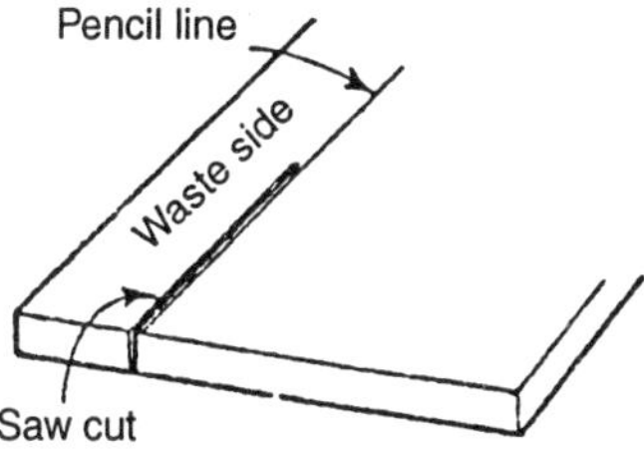

Fig. 5.44: Sawing to pencil line.

Planing Operation

This operation of truing up a piece of wood by a planer. The same is called facing and edging, and the future success of most pieces of work is dependent on this preliminary operation being properly carried out.

A properly planed surface should be perfectly straight in all directions, parallel in width and thickness, and with edges square to the face. This is achieved by carrying out the following sequence of operations. First choose what apparatus to be the best face. Next plane off the rough surface with the jack plane, taking off as little as possible, but at the same time attempting to plane the surface by taking more off the high parts. Each time the pressure is applied on the forward stroke and relieved on the return stroke.

At the start of the stroke, maximum force is applied to the handle for driving the plane. This force is balanced by pressing the other hand on the tip. For heavy work the pressure on the tip is applied with the palm of the hand, but for light work the tip is held in hand with the thumb on the top and the two centre fingers at the bottom of the sole. As planing proceeds, the material should be checked to see if it is straight across the face. This may be done with a try-square or by using the edge of the jack plane.

Chiselling

This is process of cutting a small stock of wood to produce the desired shape. To cut horizontally *with* the grain the chisel is held slightly tilted to one side and pushed forward in the direction of the grain. For roughing down to size, the bevel side is down; for the finish cut the bevel side is up. To chisel horizontally *across* the grain, the work is clamped in the vice. The blade of the chisel is grasped with the thumb and forefinger to act as a brake.

The handle is slightly lowered down and chiselled from both sides to avoid splintering the corners. With the chisel held flat the high spot in the centre is removed.

In vertical chiselling across the grain the chisel is controlled with the left hand pressing firmly on the blade of the chisel and resting on the wood.

The chisel is tilted slightly to one side to give a shearing cut. In rounding or shaping a corner, chiselling should commence at the edge of the board and work round to the end, that is, with the grain, otherwise the corner of the chisel will bite into the grain and split the wood. Firmer chisels are used for heavier work and should be hit with the mallet; in fact, the mallet should be used on all chisels and gouges whenever necessary.

A mortise and a tenon which are made in almost all wooden construction can be produced with the chiselled and mallet. A *tenon* is the projected part of a piece of wood and is cut squared and finished to the desired size to suit the corresponding *mortise,* which is recess.

Boring

To process the round holes in the wood is called boring. This boring can be done straight or inclined to suit the type of work. While boring, the work is firmly secured in a holding tool in order to avoid production of an eccentric hole. Small holes can be made by using a bradawl and gimlet, but large holes require braces and bits or drills. These bits and drills should be turned constantly in one direction and withdrawn at intervals, to remove the waste core, by turning in the apposite direction and exerting an upward pull. When using the brace and twist bit to bore rather deep holes, the direction of the bit should be carefully checked at the start. The bit should be guided by sighting either with the try square or small straight edges. The operator must stand in a suitable position while boring in order to do this sighting. Correct marking and location of the centre are also very important to produce a hole of the correct size.

The Grooving

This is a term which is used with the term tonguing. These are operations of making grooves and tongues that are usually cut on the edges of planks and boards to join them together to form big boards of large width. A groove is a channel cut to any shape, and a tongue is the corresponding projection formed to fit into it. Actual application of grooves and tongues can very well be seen in drawing boards, floor boards, wall partitions and in other articles where considerably large sizes are needed. The grove is cut with a plough plane, and a tongue with tonguing plane or a moulding plane.

The Rebating

Rebating is the process of making a rebate taken out of the edge of a piece of wood. Examples of the rebate are: that part of the door frame in which the door fits, or the recess in a window sash which contains the glass. A rebate is made by a rebating plane. Rebates which are too wide for the rebate plane may be worked by running a few grooves with the plough plane to the correct depth, and then cleaning out the bulk of the waste with mallet and chisel.

Moulding

It is also the process of cutting concave, context and other curved surface along the length of a piece of wood. This is done by a moulding plane; fixing a cutter iron of the shape required. Moulding work is done for preparing photo-frames and other decorative works.

CARPENTRY JOINTS

There are two main constructional works like framework and carcase work. In framework, typical joints used, are the various *halving joints, mortise* and *tenon joints*, and *bridle joint*. Carcase work is characterized by box-like shapes of solid wood or laminboard. Typical joints used, are *butt* or *rubbed joint, dowel, tongue and groove,* and the *screw* and *slot joint;* other joints include *dovetail joints,* and *corner joints*. Before any joints can be attempted it is necessary to prepare the stuff. This means planing the wood to size and getting four true surfaces.

To secure the corners and intersections of framing halving joints are used. The halving, joint, also termed a *half-lap joint,* may be usefully employed in many types of framing where strength and appearance are of secondary consideration. Various forms of halving joint are shown in Fig. 5.45.

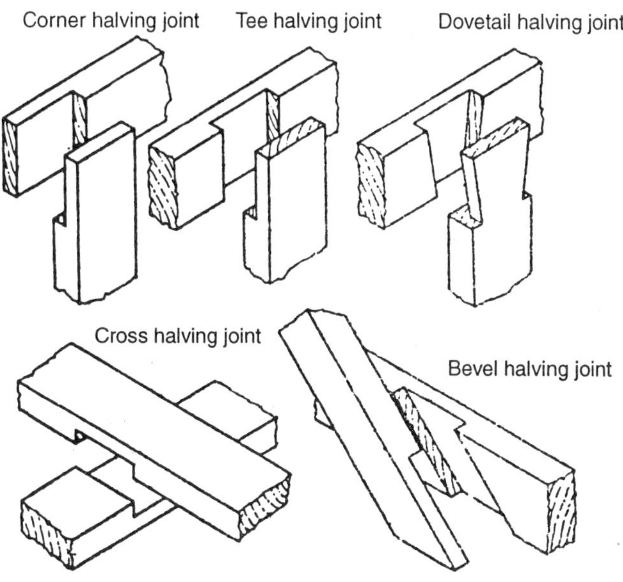

Fig. 5.45: Halving joints

Mortise and tenon joint. This family of joint is a large one and is probably the commonest used by the woodworker. It consists of a rectangular peg (tenon) fitting into a rectangular hole (mortise). Various forms of mortise and tenon joints are illustrated in Fig. 5.46. In making these joints the tenon is made by shaving only, except that in very wide ones the shoulders may be finished with a plane. After preparing the stuff the position of the tenon and mortise is squared on the wood with the pencil and then cut to prepare the pieces for making joints. Two tools have been developed solely for making the joints:

(1) mortise chisel, and (2) mortise gauge. For general framing work the width of a mortise is about one-third the

thickness of the material to be mortised, and the length should not exceed six times the width.

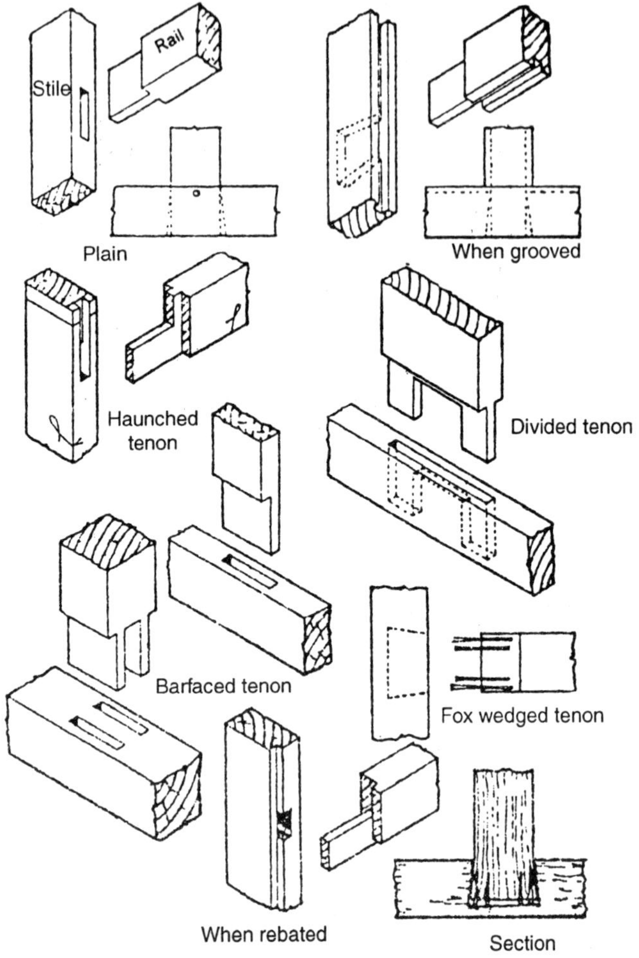

Fig. 5.46: Mortise and tenon joints

Bridle joint. It is one of the form of the reverse of the mortise and tenon joint so called open mortise and tenon. Different forms of bridle joints are shown in Fig. 5.47. The marking out is the same as for the through mortise and tenon joint, except for the placing of the cut lines. It will be noticed that there is no mortise, but two grooves; and what was a tenon now becomes a slot. The joint is often used where the members are of square or nearsquare section and thus unsuitable for making a mortise and tenon joint of good proportions.

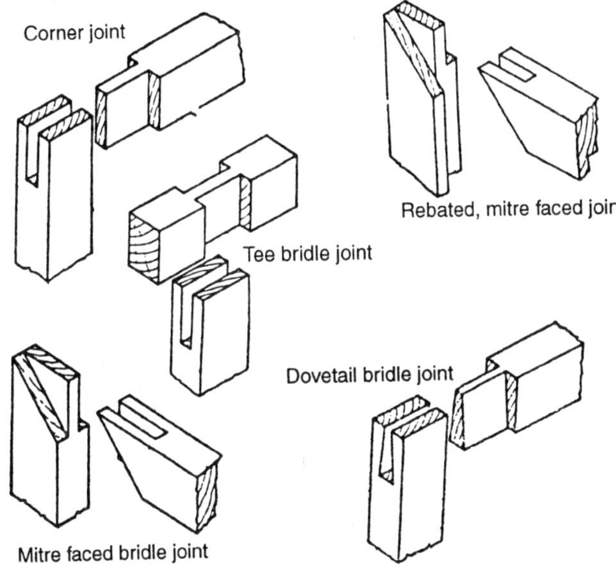

Fig. 5.47: Briddle joints

Rubber Joint. Rubbed joint is the form of edge joint which is the rubbed joint in which two true edges are joined with glue. If properly done, this joint is very strong. For stuff thicker than 25 mm, additional strength is often provided by the use of dowels or screws and slots. The rubbed joint is made by planing the two edges true with a trying plane.

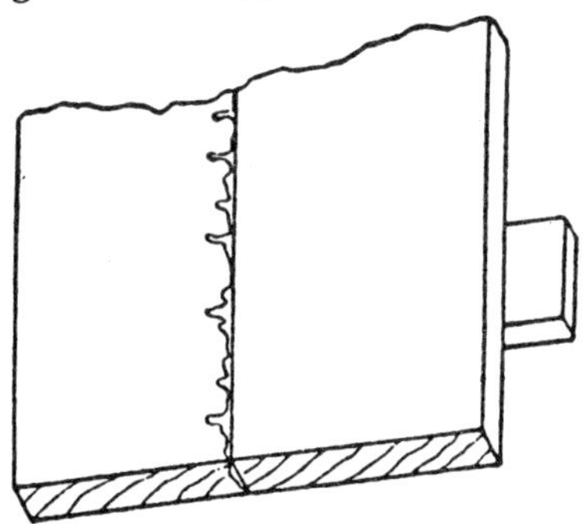

Fig. 5.48: Butt or rubbed joints

Dowel joint. Farming joint are made by dowel joint in the place of the mortise and tenon joint. This may be used to advantage in many cases, such as the joints in circular work, butt jointing two edges or positioning movable fittings. A typical dowel joint is shown in Fig. 5.49.

Tongue and groove joint. As shown in Figure 5.50, commercially machined boards for edge to edge jointing, such as drawing boards, floor boards, and match boards, are tongued and grooved. The tongues are used to provide extra support and additional gluing surface. These may be either self-tongues or loose. Self-tongues are prepared by cutting a tongue on one edge and a suitable groove on the other with the aid of a matching plane. In those tongues, both edges are trued and then plough grooved.

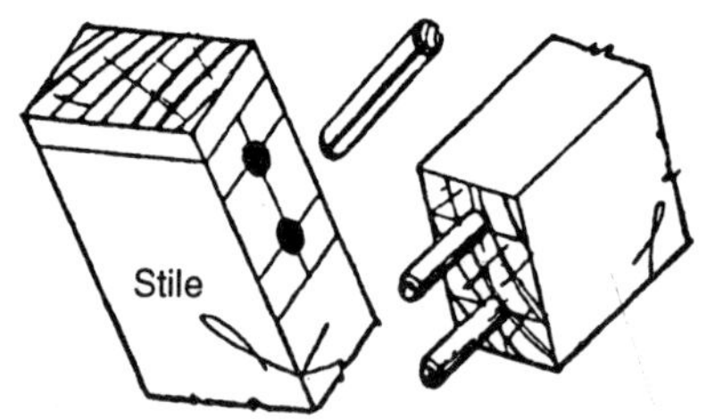

Fig. 5.49: Dowel joints

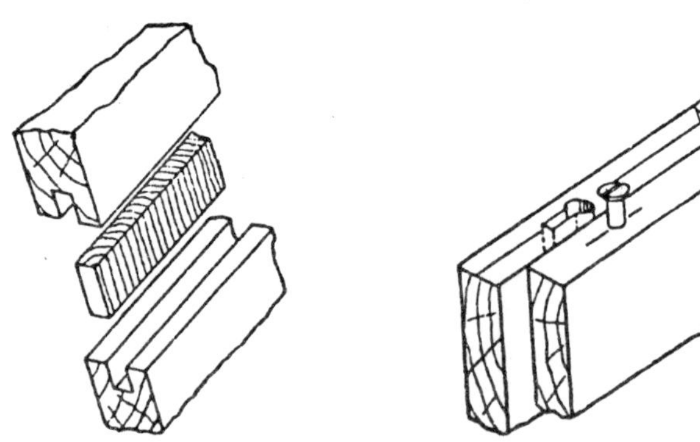

Fig. 5.50: Self tongue or loose tongue

Fig. 5.51: Screw and slot joint

Screw and slot joint. Screw and slot joint can be used as shown Fig. 5.51, where two pieces are to be secured secretly and glue may not be advisable. One piece carries the screw, while the other piece has slots cut in it to take first the head and secondly the body of the screw.

Dovetail joint. This joint is probably the strongest of all corner joints. It was primarily a joint intended to take a strain in one direction, but it has several variations and many applications, particularly in making box or carcase-like constructions – form small boxes to large pieces of furniture. Various forms of dovetail joints are shown in Fig. 5.52.

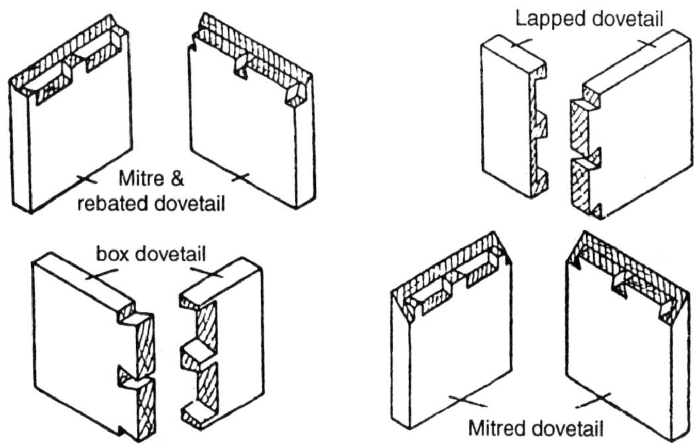

Fig. 5.52: Dovetail joints

One method is to cut the pins first and then mark the sockets from them, and the other is to cut the sockets first.

In the first method the marking out may be done neater, but each piece must be delt with separately, whereas by cutting the socket first, a number of pieces may be delt with in one operation. This is perhaps the best method, as the sockets are easier to cut straight and there is saving of time.

Corner joint. The joints of joining angles or corners together other than by dovetailing.

CARPENTRY MACHINES

Carpentary requires the use of some power-driven machines, particularly where large scale production is to be obtained. The size and capacity of the machines used depend on the size of the general run of the work to be done.

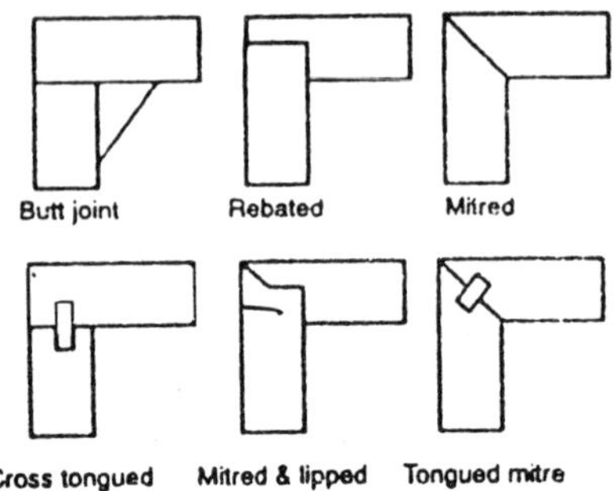

Fig. 5.53: Corner joints

Machines chosen for carpentry shop must be well built, and their accuracy should be dependable. They should be well guarded to protect the worker from the hazards of operation. The machines commonly used are briefly described below.

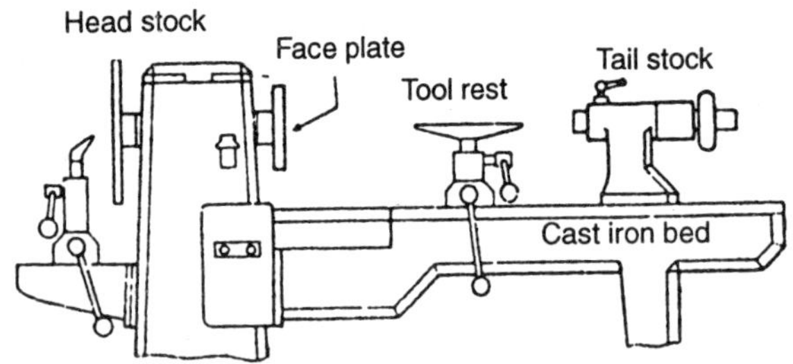

Fig. 5.54: Wood turning lathe

The Wood Working Lathe

This is the most important machines used in a carpentry shop. This is employed primarily for turning jobs in making cylindrical parts. However, by suitably manipulating the tools, tapers, radii, and other irregular shapes can also be easily turned.

It resembles the 'engine lathe' most frequently used in the machine shop, and consists of a cast iron bed, a head stock, tail stock, tool rest, live and dead centres, and a speed control device (Fig. 5.54).

The drive, in modern lathes, is individual motor driven; and a cone pully on the head stock spindle is connected by a belt to a cone pully on the motor shaft. In practice, the workpiece is either clamped between two centres or on a face plate. Long jobs are held between the centres and turned with the help of gouge, skew chisel, parting tool, etc.

Generally, the lathe is supplied together with a number of accessories for making it useful for a variety of jobs. The size of a woodworking lathe, as in the engine lathe, is usually specified in terms of the so-called "swing" of the lathe and the maximum distance between centres.

The Circular Saw

In carpentry shop the circular saw is the second most important single machine in a carpentry shop is the circler saw. It can be used for ripping, cross cutting, mitering, bevelling, rabbeting, and grooving. Although there are many types of circular saws such as universal saw, variety saw, bench saw, the basic working parts are common to all. Each has a flat surface or table upon which the work rests while being cut, a circular cutting blade, cut-off guide, and a ripping fence that acts as a guide while sawing along the grains of the wood.

The circular saw usually has provisions for tilting the table upto an angle of 45° to enable the machine to cut at different angles required during mitering, levelling etc. The size of a circular saw is determined by the diameter of the saw blade.

The Band Saw

This is designed to cut wood by means of an endless metal saw band that travels over the rims of two or more rotating wheels. Other parts of a band saw are frame, table, saw guides, saw tensioning arrangement, etc. Although the number of operation that can be performed on a band saw is less than those of a circular saw, it is most useful for making curved or irregular cuts in wood.

There are two models of the band saw as vertical and horizontal. In the former, two wheels are arranged side by side and the table is mounted underneath. In the latter model, illustrated in Fig. 5.55, the wheels are arranged one above the other in a vertical plane below the table and the band passes through the table.

As in the case of the circular table, angular cuts are obtained by tilting the saw table. The size of the band saw is specified as the distance from the saw band to the inner side of the frame. This distance is roughly equal to the diameter of the wheels.

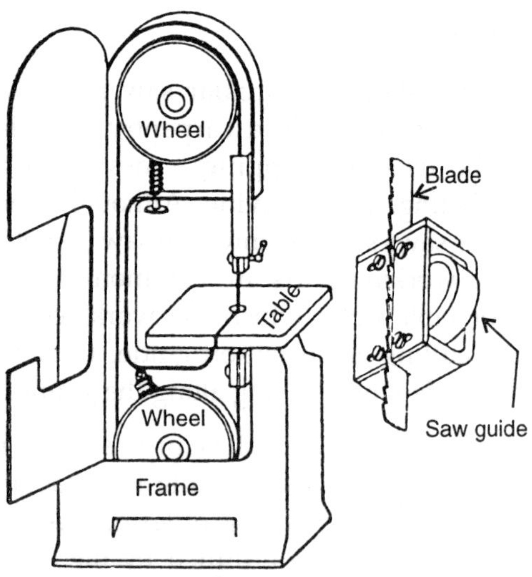

Fig. 5.55: Band saw and guide unit

Scroll Saw

The scroll saw is used for making intricate and irregular cuts on small jobs. On thin wooden pieces, jig saw can cut in a curvilinear path. The machine is actually a type of band saw of much smaller size and specially adapted to irregular work.

The scroll saw consists of a base, frame, table, upper and lower chucks, guide assembly, and blade. Chucks hold the blade with its teeth pointing downward. The blade resembles a hand hacksaw blade in regard to its shape. The blade reciprocates vertically up and down and shapes the wood. The table of the jig saw can be tilted for angular work. The special feature of the saw is that it can be used to cut inside curves. This is specified by its blade-to-arm distance.

Jointer

For planing straight edges and surfaces of boards, a jointer is used. In practice, it performs the work of a hand planer and is capable of producing a true surface with sufficient accuracy and speed.

A jointer has a frame, table, feed rollers, revolving head fitted with two or three cutter knives. With the help of feed rolls, the plank is fed to the cutter head which removes the wooden chips as the board advances and makes its surface smooth and plane. By means of an adjustable fence, the jointer can also be used for angular and level cuts. A jointer is specified by the length of the cutting blade.

Planer

It is designed for planing heavy stock at a faster rate. The boards to be planed are fed by means of feed rolls along a table against a revolving cutter head. The cutter head is mounted on an overhead shaft which is adjustable for regulating the depth of cut. The table of the planer is much wider and longer than that of a jointer for accommodating large and heavy stock.

Thmortiser

This is a square slot cut in the direction of depth for the purpose of making a mortise and tenon joint in a wooden piece. The mortising machine is used for cutting mortise and tenon joints which are very laborious and time consuming operations. There are three types of mortisers, namely,

(1) hollow chisel mortiser, (2) chain mortiser, and

(3) oscillating bit mortiser.

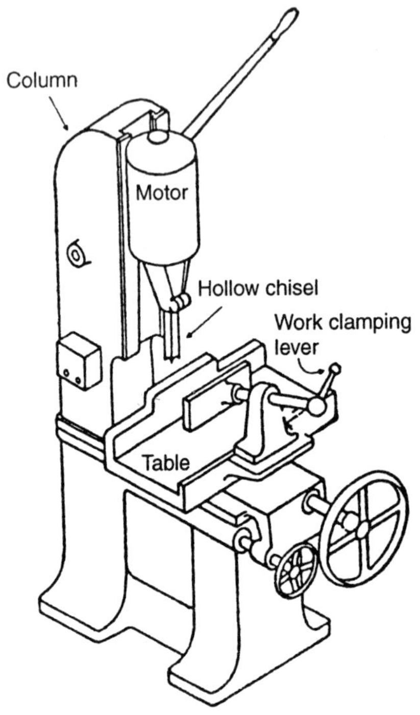

Fig. 5.56 Power mortiser

The *hollow chisel mortiser* consists of a revolving spindle carrying an auger bit at the bottom end. The auger bit rotates at a high speed inside a hollow chisel of square section. When the chisel is forced into the wood, the bit bores a square hole by the sharp end of the chisel. The auger bit and chisel thus work together and perform boring of a square hole. The depth of the mortise is regulated by means of an adjustable depth stop. The spindle is rotated by an electric motor, and tool-feed is obtained by pressing foot-lever. This is illustrated in Fig. 5.56.

The *chain mortiser* which is primarily used for making mortises in doors and windows. It carries an endless chain which has saw type teeth on its outer surface. The chain revolves around a guide bar and cuts the stock. The mortise of the desired length is produced with round bottom corresponding to the profile of the revolving chain.

The *oscillating bit mortiser* carries a oscillating router bit and produces comparatively small mortises suitable for small cabinet and chair work.

The Sanding Machines

Essentially, a sanding machine performs a sand papering job to produce a uniformly sanded surface. Three common type of sanding machines are:

1. Belt sander.
2. Spindle sander.
3. Disc sander.

Belt sander. Belt sander has an endless cloth backed abrasive belt which runs over two drums and is used for sanding and shaping flat surfaces. One of the drums is rotated by an electric motor and serves as the driver, while the other supports the belt and keeps it in proper tension. For sanding work by the abrasive belt, the workpiece is supported by an adjustable table that may be tilted to any desired angle.

Spindle and disc sanders. During operation, the disc or the spindle as the case may be, rotates and performs the work.

The Grinder

This is a small tool which is a must for all woodworking shops for sharpening and shaping various tools used in the shop. The grinder has two grinding wheels fastened on to the two ends of a rotating spindle which is driven by a small electric motor. Generally, one of the wheels is used for coarse grinding while the other for fine grinding. Sometimes one of the wheels, particularly which is softer, is provided with a wet-grinding attachment.

MANUFACTURE OF PLYWOOD

This is made of thin wooden pulps called veneers bonded together with an adhesive (synthetic or natural). The veneers are placed on each other in such a way that grain of each alternate layer crosses the adjacent layer at right angle. The plywood becomes extremely strong, stable, rigid for such arrangement of veneers. It is having high resistance to impact also.

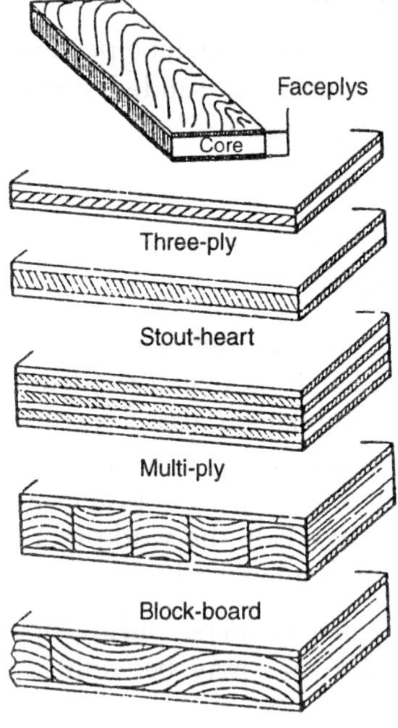

Fig. 5.57

The veneers used should not be less than three (minimum) and may be as many as 15. The art of veneer peeling and slicing has been refined with minimum thickness of 0.2 mm and maximum of 6 mm now possible. Fig. 5.57 shows some varieties of plywood.

Boards with more that three laminates are known as *multiply*. Standard plywood may have thickness ranging from 3 mm to 25 mm. But thinner or thicker plywood can be manufactured for special purposes.

Block boards are manufactured in similar length and width as plywood and in thickness of 12, 16, 18, 24 and 25 mm. The core consists of 25 mm wide wooden strips and outer faces are good quality veneers of thinner sections (1-3 mm). In *batten board* the core is more than 40mm wide.

Chipboard is manufactured from fine wood particles or woood wastes from other timber-processing system, bonded together with synthetic resin glue under heat and pressure. *Oriented-strand-boards* (OSB) are engineered, mat-formed panel products made of strands from small diameter round wood logs and bonded by phenol-formaldehyde under heat and pressure. They are stronger than chipboards and meet nearly all the specifications of ply-wood.

Plywood Manufacturing. There are nine steps in manufacturing of plywoods.

In the first step, as veneers are made from softwood and hardwood only, the log is first *debarked* i.e. outer bark is removed. In the next step the log is cut to appropriate length, known as *bucking*. The logs cut in length-wise (known as *blocks* are heated around 95°C using hot water bath, steam heat or hot water spray to soften the logs for *veneering*.

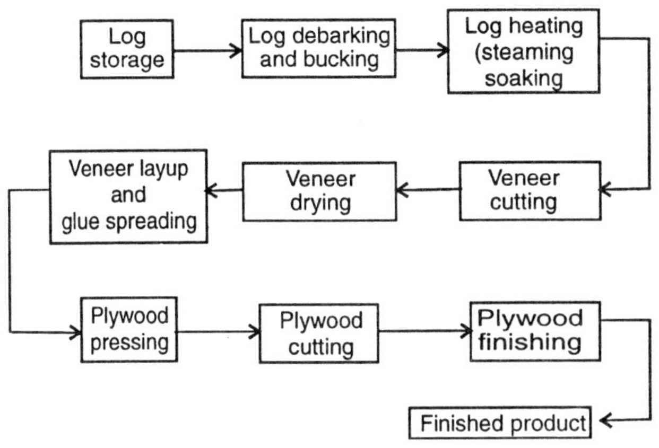

Fig. 5.58

The Veneering

Rotary cutting, the veneers are peeled from logs of good girth and clear straight grains. The veneers are peeled rotary fashion (rotary cutting) from the block by a long knife fitted to a lathe-like machine.

The knife advances at a uniform rate, peeling the veneers of constant thickness. At the end a pole like round body remains containing pith which is discarded.

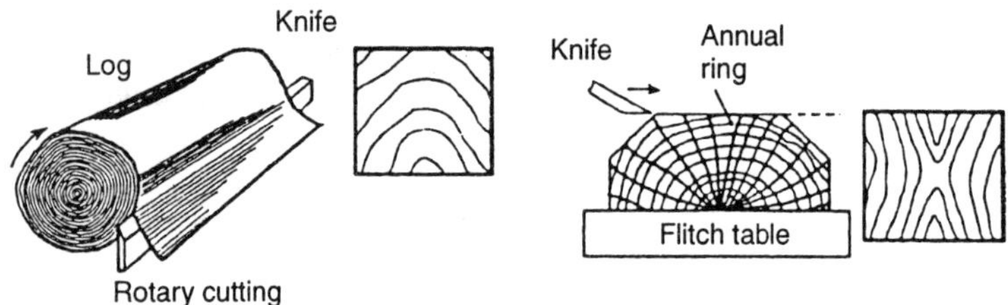

Fig. 5.59: Methods of veneering

Slicing: Slice is used for producing decorative, high quality veneer. In this method the block is first cut diametrically in two parts known as flitch. The flitch is then mounted on the flitch table and sliced. Fig. 5.59 shows the methods of veneering using *flat slicing* method.

Veneer drying and other operations: Veneers are sent to a drier to dry them in the range of 150°C to 250°C. Once the veneers reach the desired level of moisture content, they are cut to size and conveyed for *layup* operation, where a thermosetting resin is spread evenly on the veneers either by a glue spreaders on a spray system. Assembly of the plywood panels must be done such that the grains of alternate layers are at right angle. The assembly is then hot pressed, cut in size and the edges trimmed. The face and back may be sanded if needed.

MATERIALS USED IN CARPENTRY

Some of the materials utilised in carpentry operations are:

1. Nail (Fig. 5.60)
   (a) Hand-wrought–forged by hand from hot iron.
   (b) Cut nails–cut from iron strip by machine.
   (c) Cast nails–molten iron run into moulds.
2. Wood Screws (Fig. 5.61)
3. Raw plugs: Used to fix items like cabinet, mirrors, racks etc. on walls.
4. Wood preservatives.
5. Glue and adhesives.
6. Wood polish.

Wood preservatives: The wood may be deteriorated by exposure, moisture, human contact, airborne pollutants and the effects of time. Wood preservatives are used to minimise the *deterioration.*

These can be divided into two general classes:

(1) Oil borne preservatives such as creosote and petroleum solutions of pentachlorophenol, and

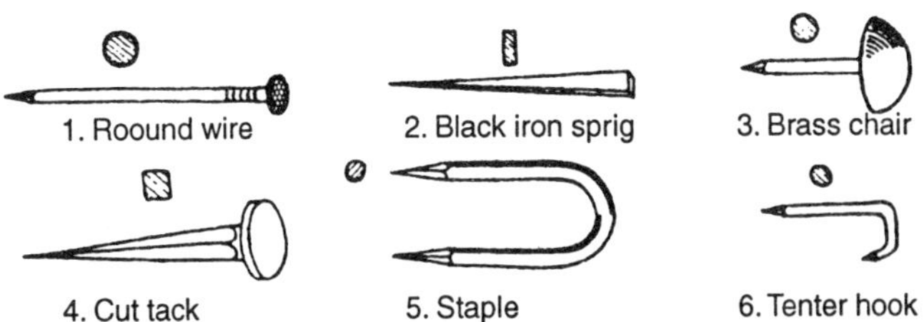

Fig. 5.60: Types of nails

(2) Water-borne preservatives that are applied as water solutions.

Coal-tar *creosote* is a black oil made by distillation of *coal tar* that is obtained after high temperature carbonation of coal. Its various advantages include:

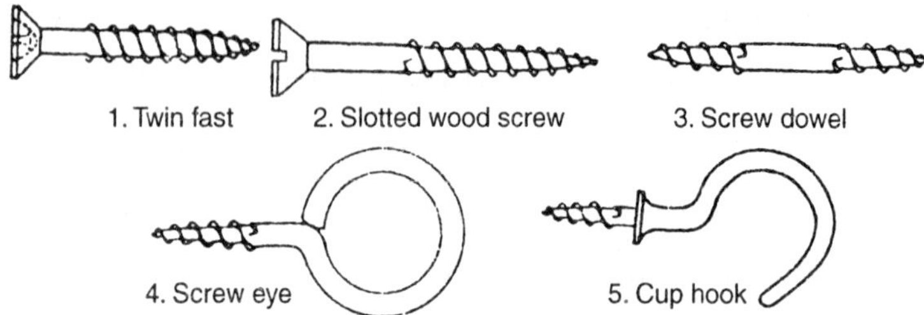

Fig. 5.61: Wood screws

(a) high toxicity to wood-destroying organism,

(b) ease in application; and

(c) relative low cost.

Glue and adhesives: Glues and adhesives are used to strengthen the bond of wooden parts after the joinery is made as follows:

1. Animal glue derived from the bones and hides of animal.
2. Casein derived from milk.
3. Synthetic resins urea-formaldehyde formulation.
4. Aerodux resorcinol-phenol-formaldehyde resins which can be used as gap-filling adhesive in wood joints, preventing fungus and insect attack.
5. Polyvinyl acetate, cold setting resin glue used for furniture assembly.

Wood polish: Wood polish is a good way of preservation, large number of wood polishes in the form of liquid and paste are available in the market. They increase the look of the wooden parts/furniture and prolong (shelf) life of the items. There are several types of polishes in the market as follows:

1. *Oil polish*: linseed oil, to be applied in numerous coats till a desirable finish.
2. *French polish*: made by dissolving shellac in methylated spirits with hardening additives.

3. *Wax polish*: applying successive coats of wax directly applied on bare wood. It is generally used for hard wood.

4. *Clear varnishes and lacquers*: may include traditional varnish like copal varnish or modern polyurethane varnish. Applied on warm and dry wood.

6

Benchwork and Fitting Shop

INTRODUCTION

Man generally do their work by hand or by the help of some specific machines. When we do ore function through the help of certain tools then tools which perform certain job or operation are called hand tools. There are certain job which do not require machines like filing, stamping, driving a screw. chipping and sawing etc. The shop where these operations are carried out is called the fitting shop and the bench where these operations are carried out is known as the fitter's bench or fitting bench.

HAMMERS

Hammer is a hard and solid tool which is made by high carbon steel. These are generally used for striking nail rivets, punches, chisels, etc. There are many parts of hammer i. e. head, pein, check, eye, face and handle. Hammers are classifed either according to thir wight or aqccording to the shape of the pein or head. The commonly used hammers in a fitting shop are:

1. Ball pein hammer
2. Cross pein hammer
3. Straight pein hammer

4. Claw hammer
5. Sledge hammer
6. Soft faced hammer
7. Mallet.

Ball Pein Hammer. The hammer which has one side spherical and other head of the ball pain is flat, it is showing in the Fig. 6.1 (a). This tool is basically used in fitting shop. A ball pein hammer is used for rounding off rivets, making round and concave indentations and flattening surfaces.

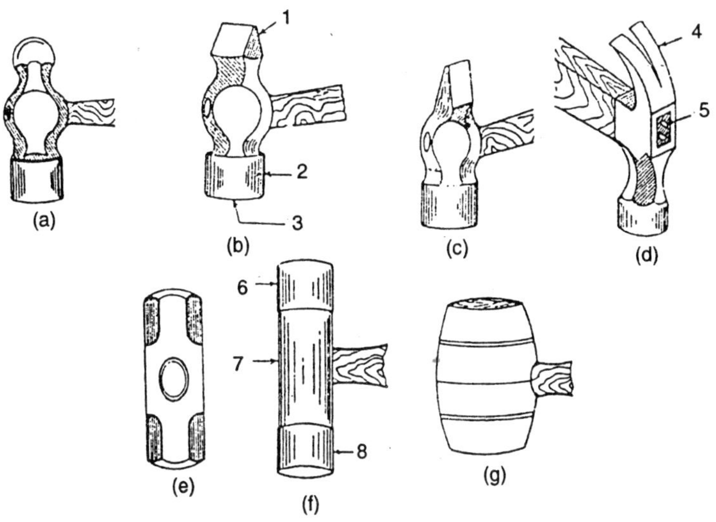

Fig. 6.1: Types of hammers (a) Ball pein hammer (b) cross pein hammer (c) straight pein hammer (d) claw hammer (e) sledge hammer (f) soft face hammer and (g) mallet. 1. The head of hammer is cross to handle 2. pole 3. face 4. clawhead 5. wedge 6. head of sof metal 7. handle and 8. head of soft metal.

Cross Pein Hammer> This type of hammer has a flat face on one end, and an edge or pein at right angles to the handle as shown in Fig. 6.1 (b). The flat edge of this hammer is used for striking and the cross pein is used for making indentations and grooves on workpieces.

Straight Pein Hammer: In such type of hammer one head of the straight pein hammer is flat and the other head has an edge (or pein) parallel to the handle as shown in Fig. 6.1 (c). The straight pein of this hammer is used for making grooves, and straight indentations on workpieces.

Claw Hammer. In this type of hammer has a cut in the centre and is bent like a sharp curve as shown in Fig. 6.1(d). This type of hammer is generally used in carpentry or pattern making shop.

Sledge Hammer> This is a heavsy hammer used mainly by blacksmiths, wood cutters or stone breakers.

Soft Face Hammer. This type of hammer as shown in Fig. 6.1 (f) is made from soft metals like copper, aluminium or lead. In a fitting bench this hammer is used for hammering finished jobs and for aligning jobs on a machine.

Mallet. Mallet hammer is made by the wood. It is used by carpenters for driving their tools and by sheet metal workers for bending thin sheets.

HACKSAW

This is a tool which is used for cutting metal rods, bars, pipe, etc. Hacksaw can be classified according to the type and shape of the frame as (a) solid frame and (b) adjustable frame. The advantage of adjustable frame hacksaws is that blades of different sizes can be fixed on them. The job to be sawed is held in a vice and the blade is moved to and fro for cutting material. The cutting operation takes place on the forward stroke and this stroke is known as cutting stroke. In the return stroke, no cutting action takes place and this stroke is called the idle stroke.

BLADE OF HACKSAW

Blades of hacksaw are made from different materials like high carbon steel, alloy steels and high speed steels. They are used in hardened and tempered condition. The vicker's hardness of hacksaw blades varies from 625 to 750 VPN. To allow the blade to move freely in the groove which it cuts, the teeth of the blade have to be set for proper frictionless movement of the blade. The teeth are set by leaving one tooth straight, bending the second towards left, the third towards right and so on as shown in Fig. 6.3 (b).

Different hacksaw blades have different number of teeth ranging from 5 to 15 per centimetre. Blades having lesser number of teeth per cm are used for cutting soft materials like aluminium, brass and bronze. Blades having larger number of teeth per cm are used for cutting hard materials like steel and cast iron.

SAWING PROCEDURE

The hacksaw blade is fixed in the hacksaw frame through

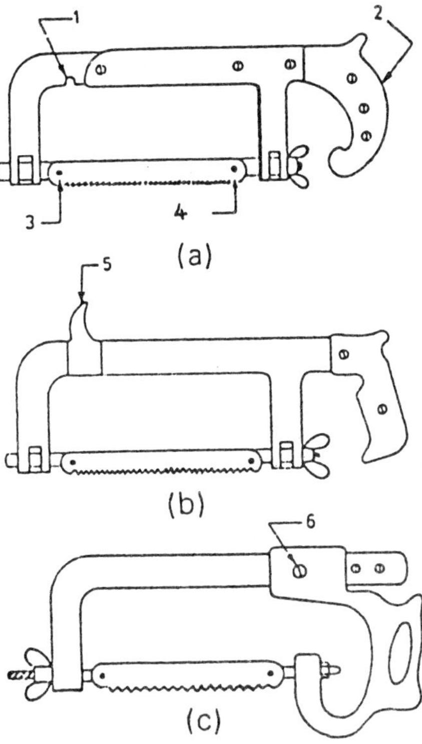

Fig. 6.2: Types of hacksaw (a) adjustable frame (pistol grip) (b) fixed frame (c) tubular adjustable frame 1. Groove 2. Pistol 3. Pin 4. Pin 5. Thumb rest rest and 6. Lock screw

this the holes a the ends fit properly on the pins of the frame. The teeth of the blade should point away from the handle (see Fig. 6.2. The wing nut should be tightened so that the blade produces a musical twang when struck lightly with the finger.

FILES

A file is basially tool which is used by the fitter. It has a large number of sharp edges or teeth that remove fine chips of materials.

Fig. 6.4. shows the parts of a file. A handle is fitted on the tang of the file. The body of the file is hardened but the tang is left soft. The edge of the file may or may not have teeth. A file having no teeth on the edge is called a safe edge file as the edge does not do any filing work. A safe edge file is used to save the adjacent face of the workpieces from being damaged.

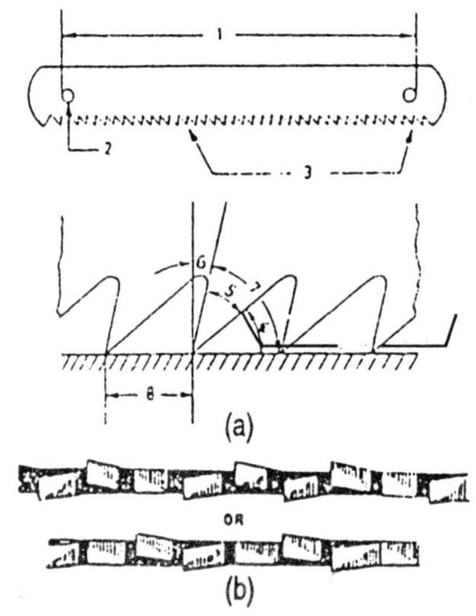

Fig. 6.3 (a): Typical view of a blade

1. Nominal length
2. Pin hole
3. Teeth
4. Rake angle
5. Lip angle
6. Clearance angle
7. Cutting anlge
8. Pitch.

(b) Setting of teeth of a hacksaw blade.

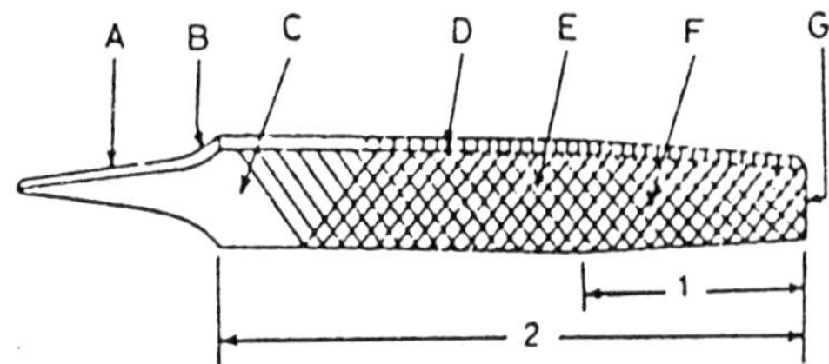

Fig. 6.4: Parts of a file

1. Taper length — 2. Length of file
A. Tang — E. Face
B. Shoulder — F. Belly
C. Heel — G. Tip or point
D. Edge.

FILE CLASSIFICATION

Files may be classifed in many ways. these are given below:

1. According to their longitudinal shape and cross section.
2. Classification based on cuts.
3. Classification based on teeth.
4. Classification based on length of the file.

Classification Based on Shape and Cross-Section

(i) *Flat File:* It consists of a long flat body with a rectangular cross-section as shown in Fig 6.5 (b). It is used for filing flat surfaces and general work. Sometimes the edge of this file is used for making rectangular grooves.

(ii) *Square File:* It has a square cross-section and is generally used for filing square holes in workpieces. Fig. 6.5 (a) shows a square file.

(iii) *Round File:* A round file consists of a circular cross section and overall shape resembles a circular solid cylinder. This file is generally used for filing round holes or concave surfaces.

Fig. 6.5 (c) shows the construction and working of a round file.

(iv) *Half Round File:* The file having a circular cross-section is called Half Round File. One face of the file is flat and other is cylindrical. The half round face is used for filing concave surfaces, while the face can be used for filing flat surfaces.

(v) *Triangular File:* It contain of a triangular cross-section of equal faces making an angle of 60° with the adjacent face. This file is used for making angular surfaces and grooves having an included angle of 60°. Small size triangular files are used for sharpening the teeth of wooden saws. Figure 6.5 (e) shows the shape of triangular file.

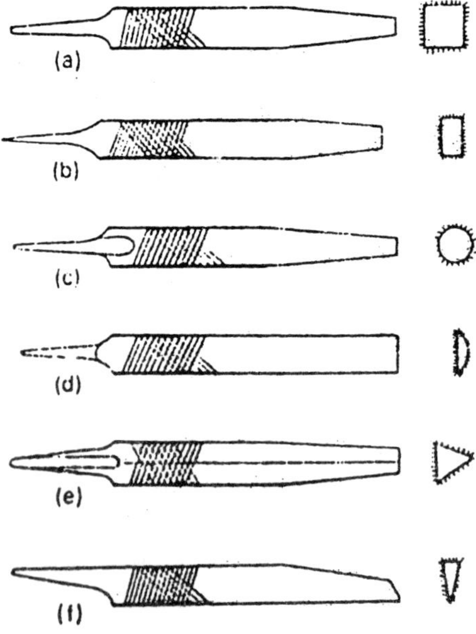

Fig. 6.5: Kinds of files
(a) Square (b) Rectangular
(c) Round (d) Half round
(e) Triangular and (f) Knife edge

(vi) *Knife Edge File:* File having similar cross-section like a knife as showing in the Fig.6.5 (f) is knwon as Knife Edge File. This file is used for making sharp and deep grooves and for filing the corners of the workpiece.

Other types of files used in fitting are the hand file, warding file, pillar file, mill file, barrette file, and rat-tail file.

CLASSIFICATIN BASED ON TEETH

Files can be classified on the basis of their teeth as bastard file, second cut file, rough file, smoth file, etc. The pitch of the file is the spece between two consecutive teeth. The pitch of the file plays an important part in the selection of files. According to their pitch files are classified as:

1.	Rough file	8 teeth per cm
2.	Coarse middle file	10 teeth per cm
3.	Bastard file	12 teeth per cm
4.	Second cut file	16 teeth per cm
5.	Smooth file	20 to 24 teeth per cm
6.	Dead smooth	40 or more teeth per cm.

In filing, the job is held in a vice at such a height that the top of the job is at level with the worker's elbow. Hold the handle in the right hand and the point of file with the left hand. The right hand exerts the force required for the cut whereas the left hand guides the cut. The left foot of the worker should be forward beneath the vice. Since forward stroke is the cutting stroke, force on the file is to be applied only on the forward stroke and the file should be lifted clear in the backward or idle stroke to avoid blunting of the teeth. The lateral as well as the forward motion of the file from right to left helps in faster cutting of the material.

FILE CARE

For making long life of the file, the proper care of files is require. Under given are some caring point to maintain the file.

1. Always store the files in a simple wooden rack at a proper distance. If files are kept in contact with each other they will get blunt.

2. When the files aren't to be used for a longer time, they should be protected from rust by lightly applying mobil oil.

3. When the files are to be used again, the oil should be removed before using them. This can be done by bwashing the files with carbon tetrachloride or caustic soda.

4. During filing operation, the teeth of the files get clogged with chips or pins. To avoid this apply chalk to the files before use.

5. Clean the file before and after use with the help of a file cleaner of file brush.

6. Use fresh files for cutting soft metals like aluminium and brass as files that have been used for cutting hard metals previously don't cut soft metals.

SAFETY RULES REGARDING HAND TOOLS

1. Do not use pliers for cutting hardened steelwire.

2. Never use a bent, dented, cracked, chipped or wornout chisel.

3. Use the right tools for the job.

4. Always keeps the tool in their proper places.

5. Never meddle or play with tools in the machine shop.

6. Never substitute a hand type socket for a power or impact socket.

7. Do not use pipe or other improvised leverage extensions on handles.

8. Plastic handles are only for comfort and don't act as insulations.

9. Don't use a pipe wrench to bend, raise or lift a pipe.
10. Ratchet mechanisms must be cleaned and lubricated perodically with light grade oil to ensure safe performance. Replace worn out parts as and when needed.
11. Use a wrench or socket whose spening fits the nut.
12. Always use approved eye protection equipment.
13. Keep the inside of socket and wrench, free from dirt and grease.
14. Always cut at right angles.

VARIOUS TYPES OF VICES

The Bench Vice (Fig. 6.6). It consists of body, sliding jaw, handle, screw, nut, spring, cotton pin and washer. The handle, screw, nut and washer are made of steel whereas sliding jaw and body are made of cast iron. Separate cast steel plates called jaw plates are fitted to the jaws by means of set screws. These jaws have serrations to hold the job firmly.

The threaded screw is made to pass through the movable jaw at the outer end. It carries a handle at the outer end and a collar at the inner end. The collar prevents the vice from coming out of the jaw when opening of the vice is carried out. It also restricts the maximum opening of the vice.

Because the bench vice is fitted on the bench, generally it is not possible to change the height of the vice or the vice itself. Thus care is needed when fitting a vice on the bench so that it is at a suitable and convenient height. For general light work the height of the top of the bench vice should be at level with the elbow of the worker. For this purpose the height of an average man is selected as height of elbow of individuals vary. For heavy work the bench should be set lower by 6 to 10 cms.

While using a vice the following precautions should always be ensured:

1. Always set the workpiece in the centre of the jaws.
2. If it is necessary to hold the job on one side of the jaws, then use a piece of steel or wood on the other side to keep the jaws parallel and free from excessive strain.
3. The Job should project out of the jaw as little as possible and must be parallel to the jaws since they act as a guide for even working.
4. Length of the handle of the vice must be sufficient to tighten the job. Never use a hammer or lever arm to tighten the workpiece.
5. Never use a vice for fitting or straightening the job.

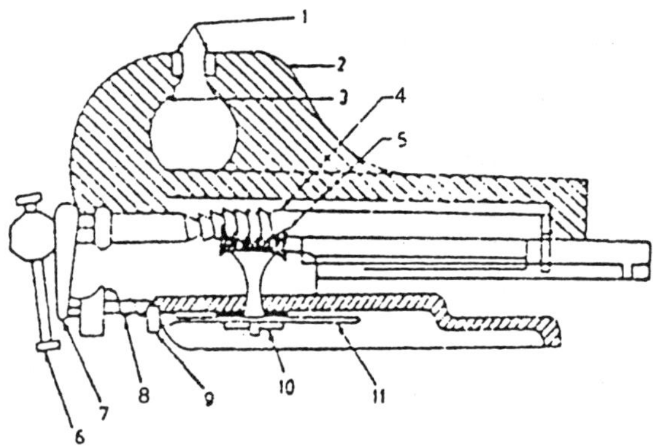

Fig. 6.6: Quick grip bench vice

1. Jaws plates
2. Fixed jaw
3. Movable jaw
4. Spindle with buttress threads
5. Half nut
6. Handle
7. Hand trigger
8. Coll spring
9. Pivot block
10. Nut bracket
11. Rocking bar

6. Fix the vice rigidly to the work bench with the help of nuts and bolts.
7. Clean the vices regularly and oil its moving parts before use for smooth working.

The Drilling Vice (Fig. 6.7)

To hold the job on the drilling machine for drilling holes in the workpiece are used the drilling vice. Drilling vice generally consists of defferent parts like handle, fixed jaw, screw, movable jaw, nut washer, etc. but its base is provided with slots through which it is fitted on the drilling machine.

Hand Vice

It is a very small machine which is used for gripping very small jobs. The commonly used hand vice is shown in Fig. 6.8.

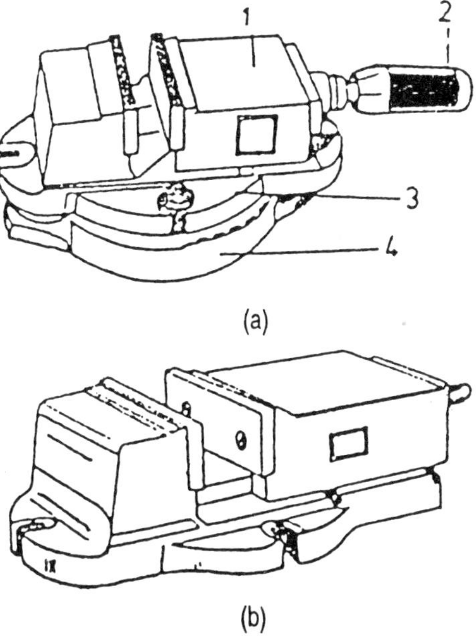

Fig. 6.7: Drilling vices

(a) Machine vice with swivel base
  1. Movable jaw
  2. Fixed handle
  3. Body fixed on base
  4. Swivel base

(b) Parallel jaw machine vice

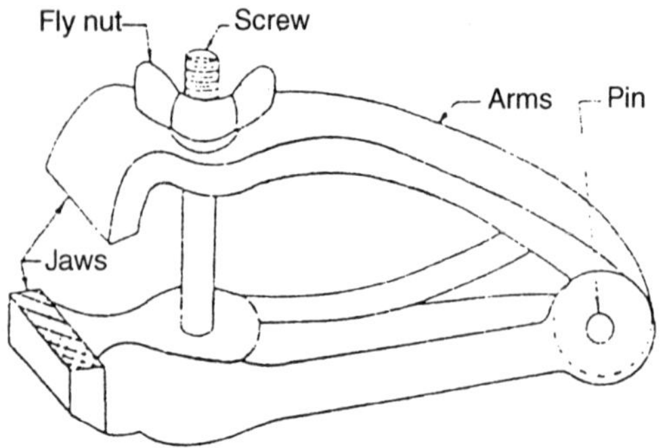

Fig. 6.8: The hand vice

SPECIAL PURPOSE FITTING VICES

Following under given are some special purpose fitting vices used in fitting shops:

(i) Toolmaker's vice

(ii) Pin vice

(iii) Pipe vice.

A Toolmaker's vice is a very simple type of vice used by a toolmaker to hold very small toools or other works. It is a hand operated vice that consists of a U-shaped body carrying a srcew in one limb. The end face of the screw and the other limb carries a metal block. Toolmaker's vice is used for carrying out small machining operations like drilling, tapping, reaming and threading.

V Block (Fig. 6.9)

This tool is of very importance used by a fitter. Basic function of V Block is to marking and drilling. The bar length is placed longitudinally in the V groove and the screw of U-clamp is tightened. It grips the rod firmly with its axis parallel to the axis of the V groove. It is used in conjunction with a U-clamp for holding round bars for marking, centre drilling, for holding in the centres of a lathe and for drilling holes.

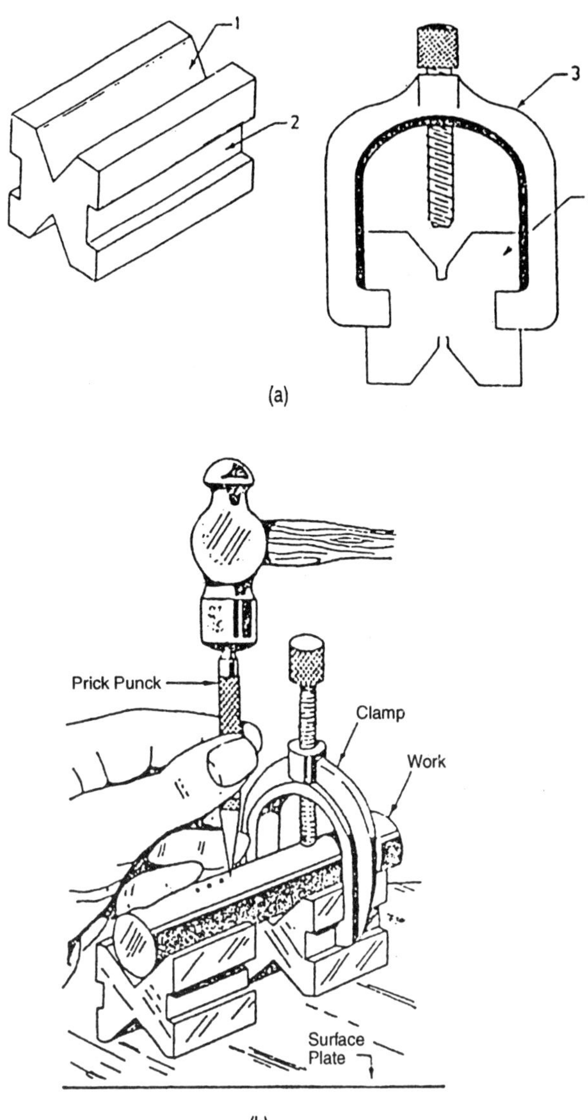

Fig. 6.9: (a) V block 1.90° V groove 2. Rectangular groove for clamp 3. Clamp 4. Vee block (b) Use of V block in layout work

C-Clamp (Fig. 6.10)

In course of assembly in a fitting shop, the C-clamp is generally used for holding the jobs. As the shape of a C-clamp used by a fitter is in the shape of a C, it is known as a C-clamp. It is used to hold different jobs together to perform fitting operations like assembly and marking.

Try Square (Fig. 6.11)

Try square is used for scribing straight lines at right angles to each other, for testing trueness of a surface and for testing mutually perpendicular surface generally it is used

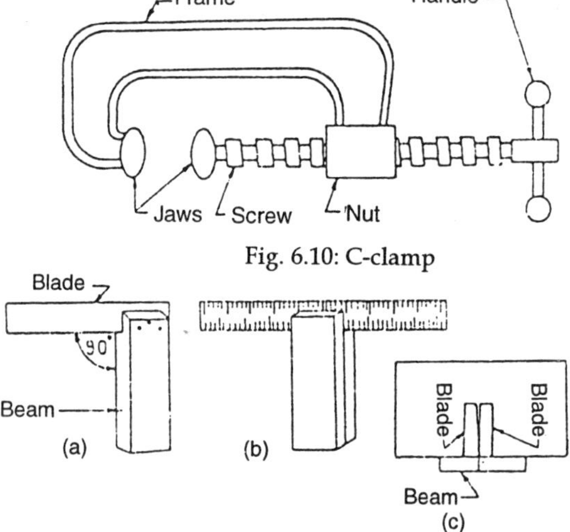

Fig. 6.10: C-clamp

Fig. 6.11: Try square (a) Solid (b) Adjustable (c) Testing accuracy of a try square

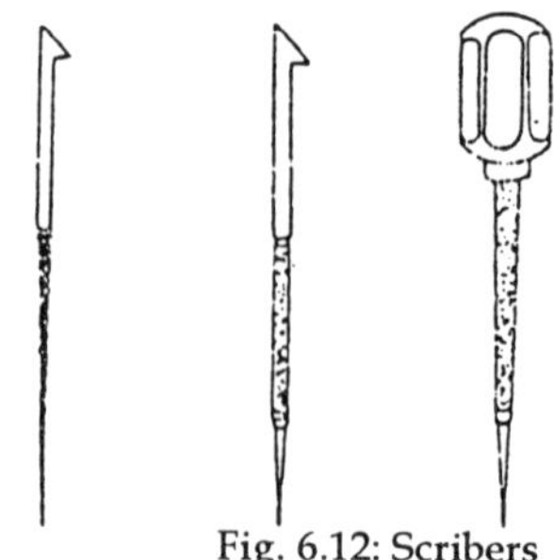

Fig. 6.12: Scribers

by a fitter in work shop. It is also called an engineer's try square is similar to a carpenter's try square but is more accurate. Try squares are made in different sizes from hardened ally steel blades. More accurate try squares have properly ground and levelled edges with a fine finish. Both inner and outer surfaces of the blade are kept true at right angles to the corresponding surfaces of the stock. The commonly used two ypes of try squares are:

1. Solid steel square
2. Double square

Solid Steel Square: Solid steel square contion a blade which is rightly fixed to the stork. It is also known as beam. Fig. 6.11(a) correctly shows the solid steel square. It is a very acuurate instrument having exact right angles, both inside and outside.

Double Square: It is also known as the movable blade try square. Double blade square is more useful for

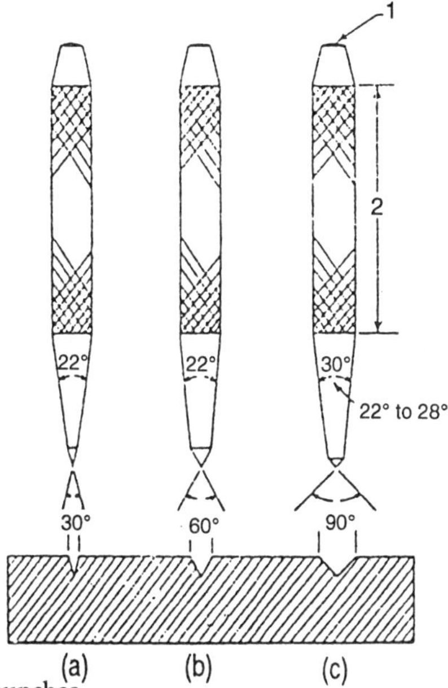

Fig. 6.13: Marking punches

(a) Prick punch with 30° point. The punch mark is not wide. This is used for thin section materials.

(b) Dot punch with 60° point. The punch mark is wider than 30° point punch mark This is used on structrural material and thick sheets.

(c) Centre punch has 90° point. It is used to provide a wide punch mark for drilling purpose.

1. Taper head for exact striking.
2. Body knurled for better grip.

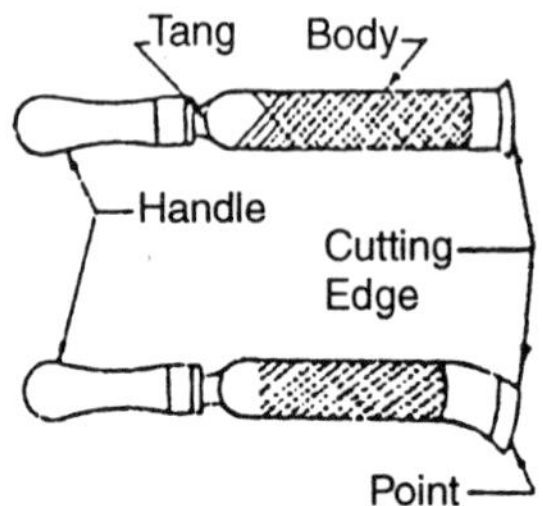

Fig. 6.14: Scrapers

more jobs than a solid squre because the sliding blade can be adjusted and interchanged with other blades. It is called a double square because it can be used for testing squareness from either seid of the blade by adjusting the beam and blade. Comparatively it is lesser accurate than a fixed steel square and shouldn't be used if a high degree of accuracy is desired.

Scribers (Fig. 6.12)

It is a piece of hard steel rod. It have a needle like point on one or both sides. It function as like a pencil to scribe lines on metals. The basic function of all scribers are same. For proper use the end of scriber must be kept sharp. If the end of the scriber gets blunt, it can be sharpened with an oil stone.

Prick Punch and Centre Punch (Fig. 6.13)

It consists of a hardened steed rod having a sharp round point with an included angle of 30°. Under given figure is showing a prick punch. It is used for making small a prick punch. It is used for making small punch marks on layout lines to preserve their location of further operations.

On other is hand a centre Punch used to mark the centre

of the holes to be drilled. In construction it is similar to a prick punch but is usually larger in size. It is made of high carbon steel which is suitably hardened and tempered. Its pointed end has an included angle of 60° or 90°.

Scrapers (Fig. 6.14)

It is a specific type of operation of removing convex spots from the machined surface and the tool used for removing complex spot is called scraper. For obtaining a fine surface. finish It is a very important hand operation employed in fitting shops. Depending upon the specific work to be performed, scrapers vary in size and shpe. Scrapers are usually made out from wornout or used old files that can't be used for further filing work. For manufacturing scrapers, the files are heated to forging temperature and forged to the rquired shape. Then further hardening and tempering operations are performed on files to get the desired hardness. The cutting edge of the scraper is ground to shampen it for removing the material. Fig. 6.14. shows the commonly used scrapers in the fitting shop. The toolmarker's scraper is knurled at one end while the other end has the cutting edge.

CHISELS

These are used for chipping work in a fitting shop. These can be classified mainly in two manners. According to the temperature of use, chisels are classified as (a) hot chisels and (b) cold chisels. According to the from of cutting edge, chisels are classified as

(i) Flat chisel, (iii) Round nose chisel, and

(ii) Cross cut chisel, (iv) Diamond point chisel.

These all are produced by forging. Hot chisels are produced by forging a bar stock of 4% tungsten. A cold chisel is produced from a bar stock of 0.9% carbon or low chromium allow steel. Usually chisels are produced from bars of octagonal or hexagonal cross section in the desired shape. The forging operation is followed by annealing, hardening and tempering to make the chisel tough to withstand sudden shocks.

SURFACE GAUGE

It is also called sribing block. It consists of a heavy flat cast iron block fitted with a vertical steel rod as shown in Fig. 6.15. On the steel rod is fitted an adjustable device carrying a knurled screw at one end and a hole for a ccriber at the other end. The scriber can be loosened or tightened with the rod by means of a knurled screw at any position by moving it up and down on the vertical rod. Surface gauge is mainly used for locating the centre of round bars held in V block by drawing straight lines and by tilting the job through different angles. Normally it is used in conjunction with either a surface plate or a marking table.

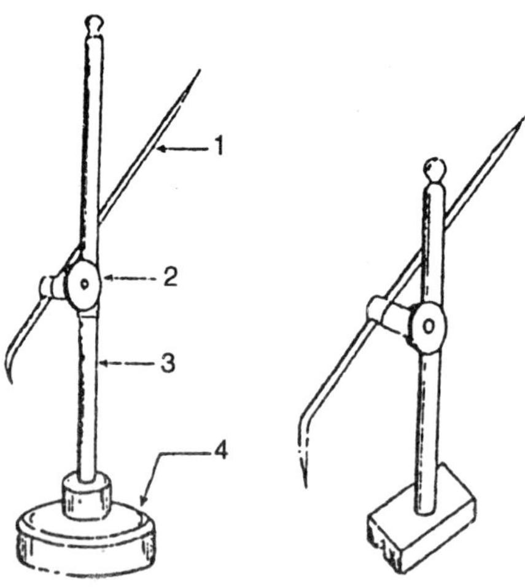

Fig. 6.15: Surface gauges with fixed spindle (ordinary surface gauges)
(a) Round base surface gauge
1. Scriber 3. Spindle
2. Scriber 4. Base
(b) Rectangular base surface gauge

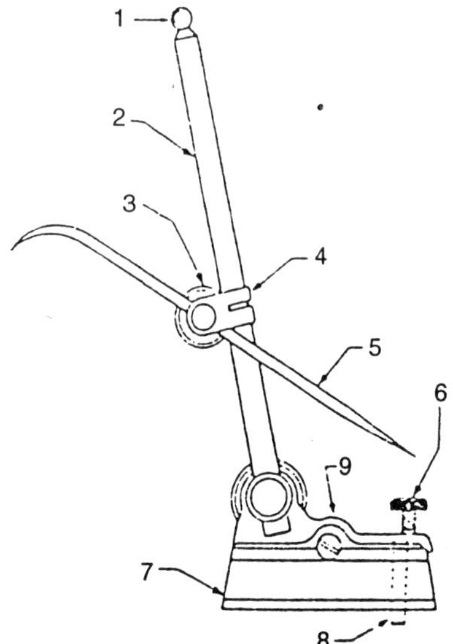

Fig. 61.16: Universal surface gauge

1. Ball head
2. Spindle
3. Scriber nut
4. Snug scriber
5. Scriber
6. Fine adjusting screw
7. Base
8. Guide pins
9. Rocker arm

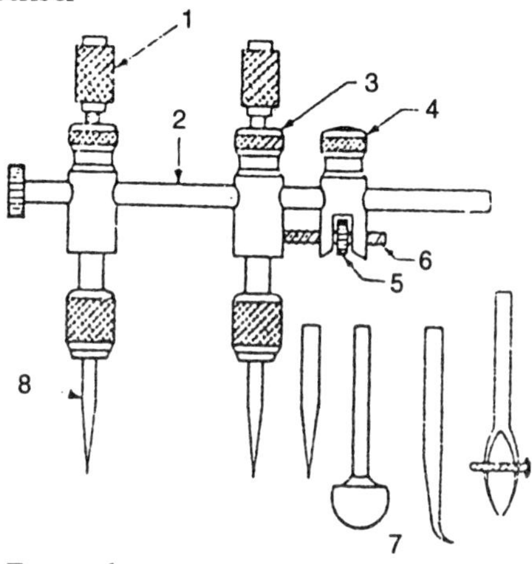

Fig. 6.17: Trammel

1. Clamping screw
2. Solid beam
3. Sliding head
4. Carrier
5. Fine adjusting nut
6. Fine adjusting screw
7. Different types of leges
8. Divider pointed leg

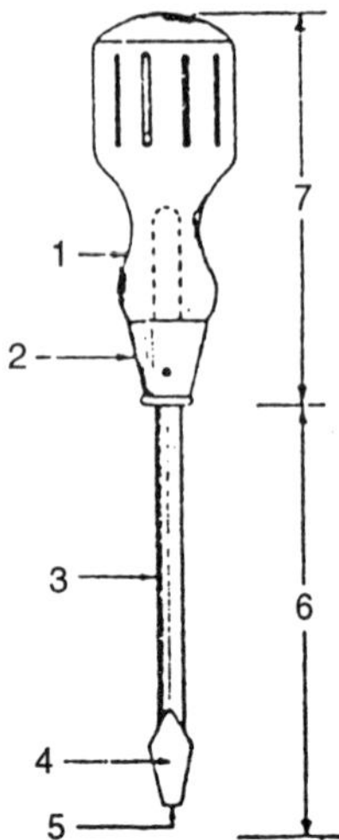

Fig. 6.18: Screw driver

1. Handle
2. Ferrule fiexs with a rivet to avoid damage to the handle
3. Shank (round)
4. Blade
5. Point
6. Length of Shank
7. Length of handle

It is an improved variety of a scribing block and is more accurate than a surface gauge. USB consist of a heavy flat cast iron or steel base carrying a spindle that can be adjusted at any angle. The scriber is mounted on the steel spindle and can be set at any angle. The base of this instrument is perfectly flat for resting on smooth surfaces. The base carries a V-groove at the bottom to enable it to rest on cylindrical objects. Two guide pins provided at the rear end of the base which can be pressed below to project down the base. The pins can be used against the edge of the surface plate or any other surface for guiding the instruments during use. The spindle can be moved in different directions by means of a adjusting screw provided at the lower end. A rocker is provided at the top of the base as shown in Fig. 6.16. The surface gauge scribes a line parallel to its base and is often used as a height gauge.

For using the USB, the instrument is placed on the surface and the spindle is set at the desired inclination. The scriber is swivelled and set at the desired height. Finally the finer adjustments are carried out with the help of a fine adjusting screw till the instrument is set to the desired height. These days this instrument is set to the desired height these days this instrument indispensable as it is easy to handle and is used or location the centres of round bars, for scribing parallel lines and camparing trueness of two similar heights.

TRAMMEL

Trammel is a divider of bigger size. It is also called a beam compass or beam trammel. It is used where a large circle having a large radius is to be marked. It consists of two adjustable vertical legs mounted on a common rod or beam as shown in Fig. 6.17. The distance between the dividers can be adjusted by opening and fastening the screws provided on the beam.

SCREW DRIVERS

A screw driver consists of a wooden or plastic handle and a steel blade having the shape shown in Fig. 6.18. Screw drivers are made to various lengths and thicknesses to suit the corresponding size of slots on the screw heads.

DRILLS

Drilling is an operation of producing circular holes of different sizes with the help of drills. In manufacture sector there are many forms of drills used for drilling holes in workshops. This is a simplest form of flat drill having a flat section at the cutting edge. It is the cheapest of all forms of drills but has the following disadvantages:

1. It isn't suitable for producing deep holes.
2. Spoilage of cutting edge is too often due to presence of the metal chips inside the drilled holes.
3. It can't be used at high speeds.
4. It isn't accurate.

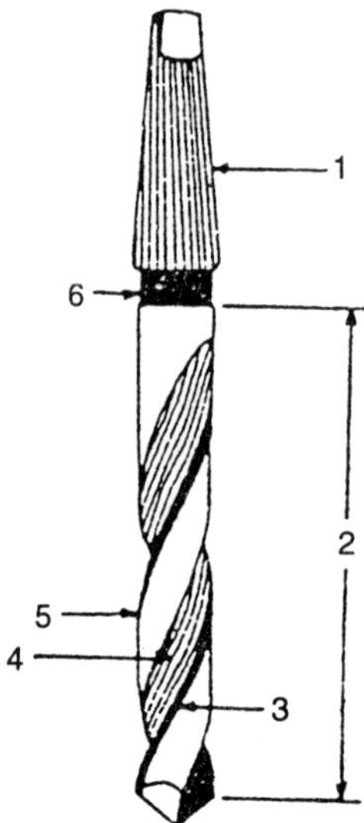

Fig. 6.19: Twist drill

1. Tang
2. Length of flutes
3. Margin
4. Flute
5. Land
6. Neck

Twist drills are the generally used drills for drilling holes in workpieces. The shank is held in the machine spindle or drill chuck. Drills are produced in taper shanks or parallel shanks. A parallel shank is provied on drills up to 12.7 mm diameter whereas taper shank is provided on drills of higher diameters. Figure. 6.19 is showing the different terms applied to twist drill. Taper shank drills are provided with a tang at the end of the shank to ensure a positive grip as this tang fits perfectly in the slot provided in the socket and also helps in easy removal of the drill from the socket. Twist drills have the advantages which are given below:

1. The chips of the metal are automatically driven out through the holes in the spiral flute.
2. The life of cutting edges is more.
3. Heavy feeds and speeds can be employed safely.
4. Power consumption by these drills is lesser as compared to other types of drills.

SPANNERS

This is a type of hand tool which generally used in the fitting shop for holding and rotating of nuts and bolts. There are three types of spanners which are commonly used in workshop.

1. Open type spanner
2. Box type spanner
3. Adjustable type spanner

Two types spanners (a) single ended and (b) double ended are used in the fitting shop as shown in Fig. 6.20.

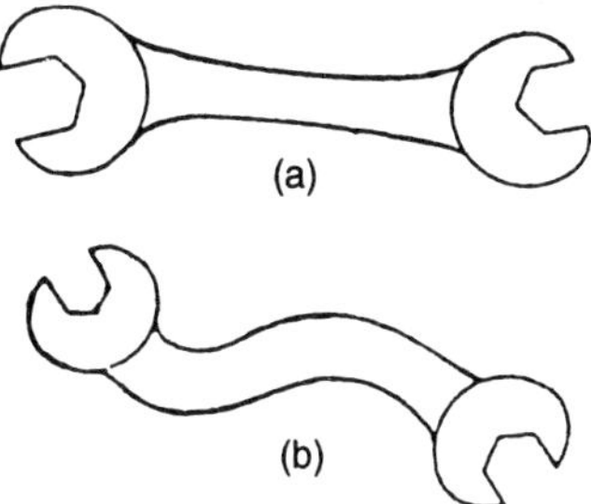

Fig. 6.20: Spanners

(a) Open-end (b) 'S' wrench

The Box type spanners are also available in two form (a) single ended and (b) double ended. These spanners do not have an open jaw but have closed ends.

An adjustable spanners consists of one fixed jaw and one movable jaw. It is provided with an adjusting screw that helps in adjustment of space between jaws to any desired value. Due to variation in opening of jaws, these spanners can be used on different sized nuts and bolts.

PLIERS

These tools are used for hoding small articles which can not hold by hand. For cutting thin wires and similar articles this instrument is used. It consists of two steel arms, each having a jaw at one end. The jaw end is shaped in different forms to suit a specific purpose. Commonly used pliers are (a) combination plier, (b) cutter plier and (c) round nose plier.

TAPS

This tool is used for cutting internal threads in cylindrical holes. Taps are of two types: (a) hand operated taps and, (b) machine operated taps.

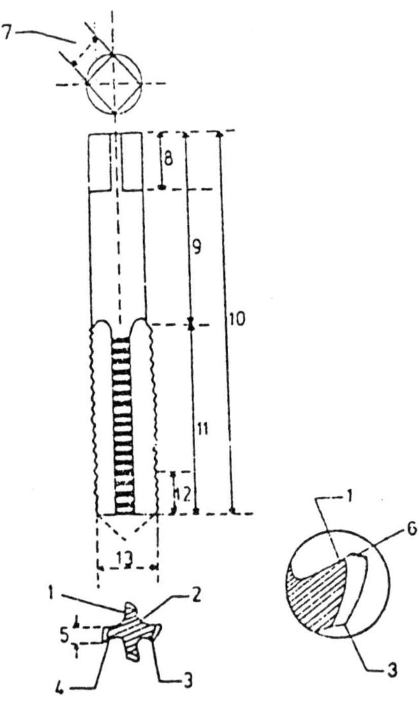

Fig. 6.21: Parts of a tap

1. Face
2. Flute
3. Heel
4. Cutting face
5. Land
6. Clearance angle or relief
7. Size of square tang
8. Length of square
9. Shank
10. Overall length
11. Length of body
12. Chamfer
13. Point dia

Hand operated taps along with a tap handle are used for cutting internal therads on workpieces. Taps are also used for cleaning damaged threads. Different parts of a tap are shown in Fig. 6.21.

A tap set consists of three taps known as taper tap, second or plug tap and finishing or bottoming tap as shown in Fig. 6.22 (a), (b) and (c) respectively. Different tap sets are required for cutting different threads on different diameters. In a tapered tap the bottom five or six thread are ground to produce a tapered surface such that the diameter of the tap becomes slightly less than the diameter of the drill hole in which the tapping operation is to be carried out. It enables the tap to enter the hole without difficulty and perform preliminary tapping operations. As the tap enter the hole each successive tooth of the tap increases the depth of the cut. After the full body of the tap has been screwed down in the hole, the taper tap is taken out of the body by providing reverse motion to the tap handle. Then the second or intermediate tap is entered into the hole and screwed down to full length. The finishing of threas is carried out by the bottoming tap by passing the body again to full length of the tap.

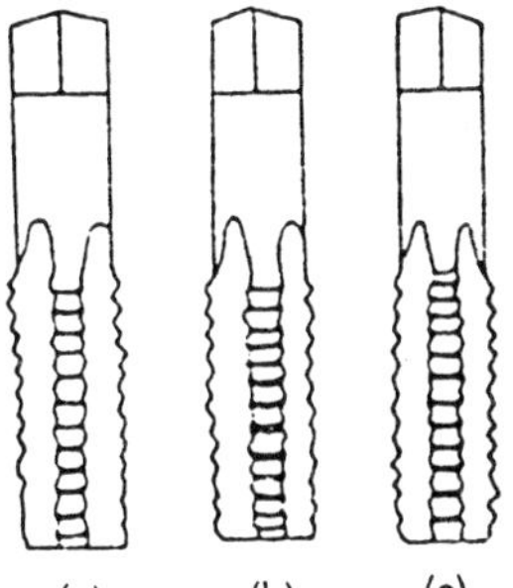

(a) (b) (c)

Fig. 6.22:Hand tap (a) 1st tap (b) 2nd tap (3) 3rd tap

Procedure of Tappings

Under given are some points which discussed the detail of procedure which is used for cutting internal threads with the help of a top set.

1. Firstly drill a hole equal to the core diameter of the thread. This diameter is known as the tapping size and is mathematically expressed as. Tapping size = Outside diameter $-2 \times$ depth of the thread

2. After drilling a hole of the required size, the taper tap is fixed in the tap wrench (or handle) and screwed into the hole. Before the start of the tapping operation the screw position must be adjusted until it stands square with the tap surface of the work.
3. Apply a little whale oil on the tap for all materials except cast iron. Lubrication improves the surface finish of threads whereas cast iron is self lubricating.
4. Start the cutting action keeping in mind that the tap is not turned countinuously but after every half turn it should be reversed slightly to clear the threads.
5. Proceed till the taper tap is through to the hole.
6. While tapping a blind hole the tap should be withdrawn time and again and metal should be cleared from the bottom of the hole.
7. Repeat the above operations with the middle tap and finishing tap to finish the hole.

DIES (Fig. 6.23)

A die is a specific tool which is used for cutting external threads on cylindrical parts i.e. studs and bolts. It contain a nut which having portions of its threads incumference cut away and shaped to providecutting edge to the remaining protion of thread. The die is held in the centre of a a pair of operating handles called stock for screwing on the bar.

The process of cutting external threads with the help of dies is very similar to tapping. Starting a threading operation with dies is a bit difficult because only one or two teeth of the die are champered and difficulty is experienced in squaring the job. When the die is working the stock should be moved forward and backward. Use lubricating oil for giving a good finish to the threads. is desirable function.

REAMERS

It is tool which used for providing a finish holes of

sufficiently good quality and accuracy. A reamer does not originate the hole in the same way as a drill, but merely imparts the previously drilled holes the necessary

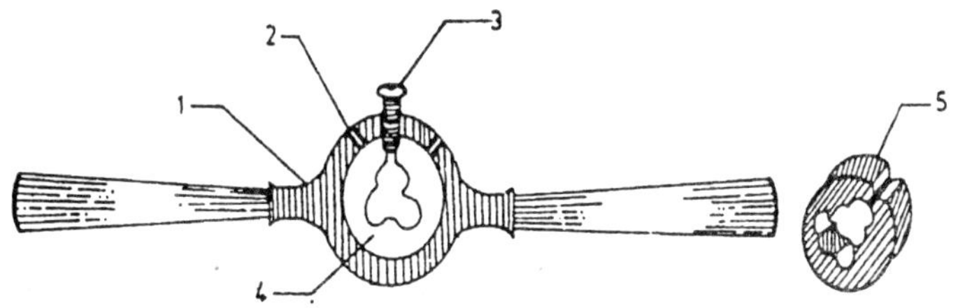

Fig. 6.23: Circular split die

1. Die holder
2. Tapped hole for screw to fix the die in position
3. Set screw
4. Circular split die
5. Isometric view of die

smoothness, roundness, parallelism and accuracy. It only finish the hole by removing very small material, but don't correct any error in position or direction of the drilled hole. The diffrent types of reamers used in workshop are (a) hand reamer, (b) machine reamer and (c) expanding reamer.

1. *Hand Reamer* consist of two main parts (a) the body and (b) head. The body consists of a number of flutes that may be straigth or helical and are tapered at the end for about one-fourth of its length. The diameter at the tip of the body is nearly 0.20 mm less than the reamer size. The shank has a slightly lesser diameter than the diameter of the flutes. While using a reamer the workpice is held in the vice and the raamer is operated by hand by holding it in the stock or top wrench. The working is similar to that of tapping. Commonly used hand reamers are shown in Fig. 6.24.

2. *Machine Reamer* It have a shorter cutting edge as compared to hand reamers. These reamers are provided with tapered shanks for holding them in

the machine. The basic machine reamer are shown in Fig. 6.25.

3. *Expanding Reamer* The depreciation in the reamers is high and their repair is expensive. When the teeth of a reamer lose their size, they are repaired (a) by reducing the size to next lower size, possible and (b) by expanding their fronts slightly by hammering eith punch (forging), rehardening and grinding the same to size. In order to avoid this expanding reamers are used. The most commonly used expanding reamer is shown in Fig. 6.24 (c).

PROCESSES OF FITTING

The fitting process in a workshop is essential operation. These are some operations carried out in a fitting section to finish the workpiece to the desired size and shape with accuracy. Following under given are some operation carried in fitting shop:

1. Marking
2. Sawing
3. Filing
4. Scraping
5. Chipping
6. Drilling
7. Internal threading (tapping)
8. External threading (dieing)
9. Reaming.

Marking: Proper and accurte marking plays a dominent role in producing a product of best accuracy. The accuracy of a component depends upon the accuracy of marking. Sufficient care is therefore needed in performing this operation to maintain quality, production and productivity of a component.

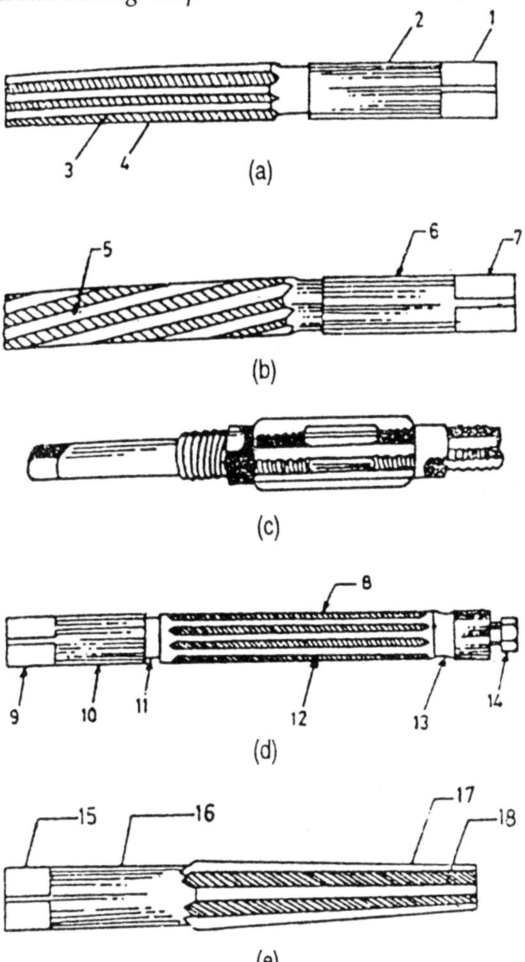

Fig. 6.24: Hand reamers

(a) Straight fluted
(b) Helical fluted
(c) Adjustable
(d) Expansion
(e) Taper straight fluted

1. Square tang
2. Shank
3. Cutting edge on land
4. Flute
5. Flute
6. Shank
7. Square tang
8. Flute
9. Square tang
10. Shank
11. Neck
12. Cutting edge
13. Neck
14. Set screw
15. Tang
16. Shank
17. Cutting edge on land
18. Flute

Marking on the workpiece is done by seeting out dimensions as given in the working drawing or by directly transferring them from a similar part. The surface to be marked is coated with read lead, french chalk or copper sulphate and allowed to dry. After this the workpiece is laid on the surface plate if it is flat or held in a V-block or angle plate. Marking of vertical lines is done by means of scribing block or height gauge.

Sawing: The process of cutting metal pieces to the expected shape and size is called sawing process. In sawing the job is held in a vice, the handle of a hacksaw in the right hand and the farther edge of the frame in the left hand. Cutting action in sawing takes place in the forward stroke and no pressure must be applied on the return or idle stroke. Do not use new blade on hard material directly and coolant must be used throughut the operation except for sawing cast iron.

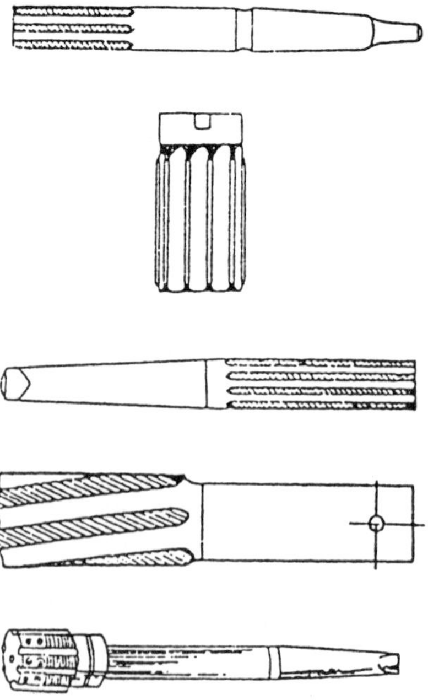

Fig. 6.25: Generally used machine reamers.

We used power saw for cutting thick materials at rapid rates. It consists of cast iron body on which is fitted adjustable vice to hold the stock. Power saw blade is fitted on the frame as shown in Fig. 6.26. While cutting on power saws proper cutting speed plays an important role. It is 40 m/min for mild steel and 55m/min for brass.

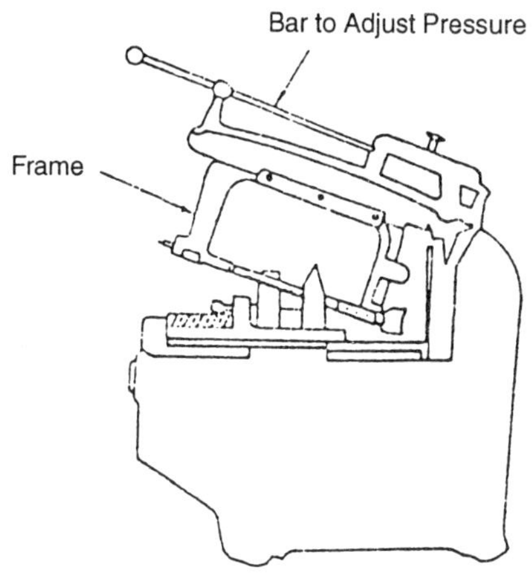

Fig. 6.26: Machine Saw

Filing: In this operation, basically the job is held in a vice. The file handle is held by right hand and by left the other side of file should be handle. The right hand applies force while the left hand only acts as a guide. The worker stands with his left foot forward and right foot behind. Normally the left foot is below the vice. The cutting action takes place in the forward direction only. In the return stroke no force should be applied on the file as no cutting action takes place in the backward stroke and application of force spoils the sharpeness of teeth.

Straight filing and draw filing are two methods of filing. In straight filing the file is pushed and pressed forward at right angles to the lenght of the job. In draw filing the file is gripped at each end and pushed forward and drawn backward along the length of the job.

Filing precautions

One should keep in mind cartain precautions while filing. these are given under:

1. Apply pressure on the file in the forward direction only.
2. Before deeping the file in the kit, clean it properly.
3. Dont't use those files on soft metals which have been used on iron and steel. It results in clogging.
4. In order to prevent clogging apply turpentine oil or paraffin while filing on aluminium piece.
5. File the chilled surface of a casting first by an old file and then with a new file.
6. Use a bastard file for rough cuts and second cuts and a smooth file for the finishing.

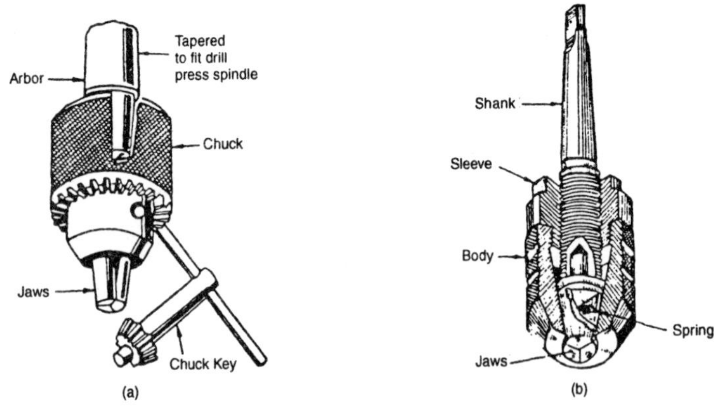

Fig. 6.27.: (a) Three jaw drill chuck
(b) Internal view of a three jaw chuck

7. When using a new file for the first time apply light pressure in the beginning.
8. For lathe work use a mill file.
9. For general work use a second cut file.
10. For filing square corners always use a safe edge file.

11. Don't use a new file for cleaning welds.
12. For filing a metal like gunmetal, brass, bronze, nickel, silver etc., use a new file.
13. Don't use a file previously used in a lathe for filing work.
14. Provide solid supports to the thin sheets before filling.

SCRAPING OPERATION

It is a specific operation which is carried out in fitting shop. The basic objective of this operation is to obtain acenrate surface that cant be obtained by filing. The operation is performed ofter filing or machining operation. Very little removal of material takes place in scraping. It is done to remove convex spots from surfaces. Scraping operation is carried out by holding the scraper in both hands and then moving it and fro.

High spots are noted on the surface when it is rubbed on a plane surface containing prussian blue or red lead. Prussian blue or red lead will stick to high surface which is noticed by turning the job upside down. High spots will contain very thin spots of red lead or prussian blue. Scraping is done on high spots till the surface becomes perfectly smooth.

CHIPPING OPERATION

The removing of metal by means of chisel is called the chipping operation chipping, the angle of cutting must be the same throughout the operation. In case the surface to the chipped is to large, it is advisable to cut the grooves throughout the surface by means of a cross cut chisel and then chip off the remaining metal. A chisel of a proper cutting angle should be used for chipping. Cutting angles of chisels for different materials has been described in previous discussion.

DRILLING OPERATION

It is a process of producting circular holes in a metal piece with the help of a drill. Produced holes may be blind or through. This process is carried out with the help of drilling machine. This machine is known as drill press.

Before carrying out drilling operation the centre of the hole to be drilled is properly located and marked by suitable means. In this operation the job is firmly held in a vice or by any other suitable means over the table of the drilling machine and the drill is fitted in the socket of the drilling spindle. The spindle machine and the drill is of the drilling press is lowered by the hand lever and it is ensured that the dead centre of the drill is in alignment with the centre of the hole. After it is ensured that the drill is in perfect alignment, the machine is switched on and the rotating drill is pressed gradually into the workpiece to produce the desired hole. Release the pressure frequently during drilling operation. While drilling thin sections, support the workpiece on a wooden block to avoid chattering and breaking of the drill. In all drilling operations the major precantion is to ensure that the job must be held rigidly.

REAMING OPRATION

Various types of reamers have been discussed earlier in the previous text. This is an accurate operation of finishing holes.

For effective working, a reamer must have proper cutting teeth. A proper stock allowance must be kept on the job to be reamed. The allowance generally provided for the johs is 0.05 to 0.10 mm. In this operation, the reamer is held in the wrench or reaming handle and forced into the hole to be reamed. The rotary motion is provided to the reamer with the help of handly by applying force to cut the metal. Differe02nt cutting teeth remove the extra metal and the hole is finished to size.

7

Welding Shop

WELDING

The art of welding is so ancient. It is old as like as any other process for working of metals. In ancient period generally blacksmith used this process for making beautiful creations. Fire welding, in which the surfaces to be joined are heated to a pasty state and then hammered together, is still practised to the limited extent necessary during the hand forging of small numbers of articles, but in all other cases fusion or electrical methods are employed.

Developments in the sphere of design, and in methods of production, have made use of the possibilities of welding to the full, so that it now stands as an important process, closely allied to other branches of workshop production.

Modern methods of welding may be classified under two broad headings:

(*a*) First is, where the parts to be joined are made solid by being melted and fused together, often with the addition of a third metal to make up the joint;

(*b*) Second is, where the parts are pressed together and brought to the temperature necessary to create the pasty state at which welding takes place.

In addition, the methods classified under (*a*) are used for depositing metal to make up for wear, to fill up cracks, to provide surfaces of another metal, and so on.

Fusion welding

In the situation when parts of a metal are collectively joined by fusion, the method is similar to a casting process because the surface being welded are heated until they melt and run together. If a filler is added, this is melted along with the metals being joined and run into the space which it is neccessary to fill up. Since the melting temperature of mild steel is about 1200°C, very high local temperatures are necessary to melt the metal at the point where the weld is to take place. The two chief methods of obtaining such temperatures are the oxy-acetylene flame and the electric arc.

Oxy-acetylene welding

There are two methods of oxy-acetylene working. One is high-pressure and second is low pressure system. When oxygen and aectylene are mixed in suitable proportions, they give a flame which in its hottest region attains a temperature of approximalety 3200°C. In the high-pressure system, both oxygen and acetylene are used from high-pressure cylinders. Oxygen cylinders are charged to a pressure of 120 atmospheres (approx. 125 bars), but acetylene, owing to its explosive tendency when compressed, is supplied in cylinders in the form known as 'dissolved acetylene. The cylinders are filled with a porous material in the four of a charcool cement and charged with acetone. This liquid has the property of absorbing 25 times its own volume of acetylene for each atmosphere of pressure applied. The pressure of the acetylene is usually about 15 to 16 bars and cylinder sizes vary from 2 to $10m^3$.

In the system of low-pressure the acetylene is generated in a low-pressuse generator by the action of water on calcium carbide, and than after it is supplied to the blowpipe at low-pressure from a gas-holder incorporated in the generator. When this method is used the acetylene has to be cleaned by passing through a purifier, and to prevent the possibility of an explosion by oxygen or air blowing back and entering the generating plant, a back pressure valve must be introduced between the blowpipe and the gas-holder. The low-pressure system is now not much used in Great Britain, but is still in use in isolated areas where high-pressure gas cylinders are not readily available, e.g. in desert oil-fields.

The pressure regulator. Before the high-pressure gas from the cylinders may be used in the blowpipe, its presure must be reduced, and the supply from the cylinder to the blowpipe is taken through a reducing valve which screws into the cylinder outlet. The pressure of the gas being fed to the blowpipe is controlled by a spring-loaded diaphragm, variations of pressure necesary for different sized nozzles being controlled by a graduated adjusting screw which varies the compression of the spring (Fig.7.1).

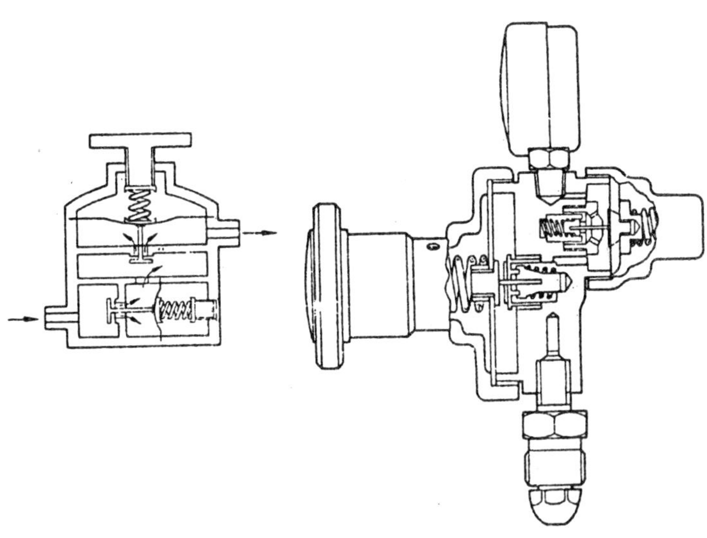

Fig. 7.1. Pressure regulator

The blowpipe. Tubes from the oxygen and acetylene supplies are carried to the blowpipe, the supply of each gas being controlled by a regulating valve. On the high-pressure system the gases pass to a mixing chamber and from there through the head of the blowpipe to the nozzle (or tip). The mixiing chamber function is to ensure that the gases are mixed in correct proportions (approximately equal) before being burned at the tip.

In the low-pressure blowpipe the oxygen is led through an injector nozzle inside the body of the pipe and in its passage draws the low-pressure acetylene into the mixing chamber, giving it the necessary velocity to preserve a steady flame.

A selection of tips is available for the blow pipe to vary the size of the flam ie heat supply necessary to weld varying thickness of metal. Each of these is marked showing its gas consumption in litres per hour, and a table supplied with the blowpipe shows which tip should be used to weld any required thickness of metal. The delivery pressure from the regulatror must be varied acording to the size of tip used, and instructions are supplied for obtaining the correct conditions.

Diagrams of blowpipes are shown in Fig. 7.2.

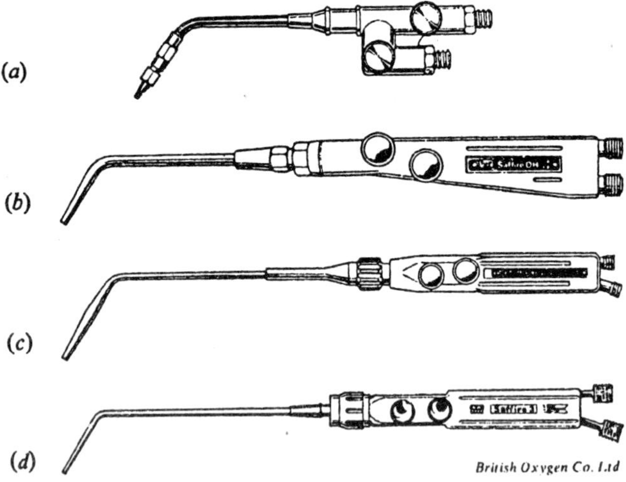

Fig. 7.2: High-pressure welding blowpipes: (a) for welding light sheet metal up to 1.5 mm, (b) and (c) for welding steel up to 8 mm, (d) for welding steel sections up to 25 mm+

The flame. For reliable welding the appropriate adjustment of the flame is crucial. The weld will be oxidise if there is an excess of oxygen. This will result in a large grain size and increased brittleness with lower strength and elongation. A carburising flame is one in which there is an excess of acetylene (C_2H_2). When welding steel this will tend to give the steel in the weld a higher carbon content than the parent metal, resulting in a hard and brittle weld. For most welding operations a neutral flame must be desired, and this is most easily judged by commencing with the carburising flame as shown in Fig. 7.3. (a), and gradually reducing the acetylene supply until the large cone just fades into the smaller one to give the neutral flame shown in (b). The oxidising flame has a still smaller cone as shown in (c). The temperature at the hottest part of the oxy-acetylene flame (just outside the cone) is approximately 3200°C.

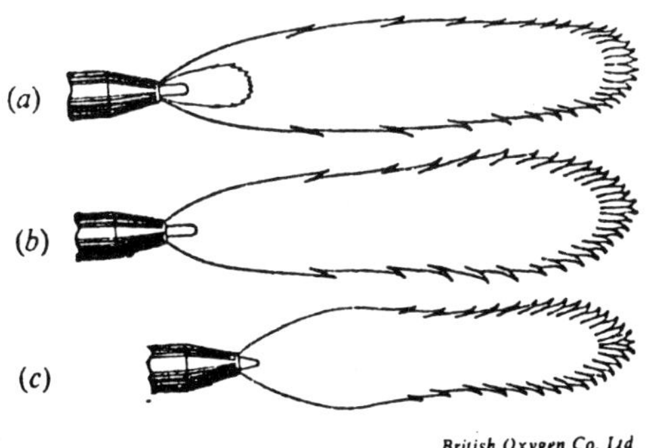

Fig. 7.3: Flame adjustment for welding. (a) Carburising flame (excess acetylene) – small excess of acetylene is necessary for Stelliting, Lindewelding, etc. (b) Neutral flame (equal quantities oxygen and acetylene) for steel, stainless steel, cast iron, copper, aluminium, etc. (c) Oxidising flame (excess oxygen), necessary for welding brass.

Welding practicing

During the last few decades the weldings operation generally used extensively in manufacturing sector. There is lot of advantages of welding in workshop operation. For instance it save the time and cost of the operations. This is a crucial and great refinement over the previous methods of production. Previously used method was generally based on orthodox principles and procedure. With the space at our disposal we cannot do more than treat very briefly the more common examples of welding practice on steel and iron, but the reader should remember that the practice has been developed so that most metals are now within its scope. A study of its technique offers most interesting examples in construction, science and metallurgy.

Butt joints. The butt joints execution is used extensively in the welding practice. Throngh the execution of butt joints plater or bar to be joined being ary thickness from 0.5 mm opwards. While functioning these work any one have choice between two methods, these two methods are: (*a*) the *leftward* (or forward), and (*b*) the *rightward* (or backward) method. In the leftward method the weld is made working from, right to left while the reverse direction is followed for rightward working. In each case the filler rod, being held in the left hand, is carried to the left of the blowpipe (Fig. 7.4).

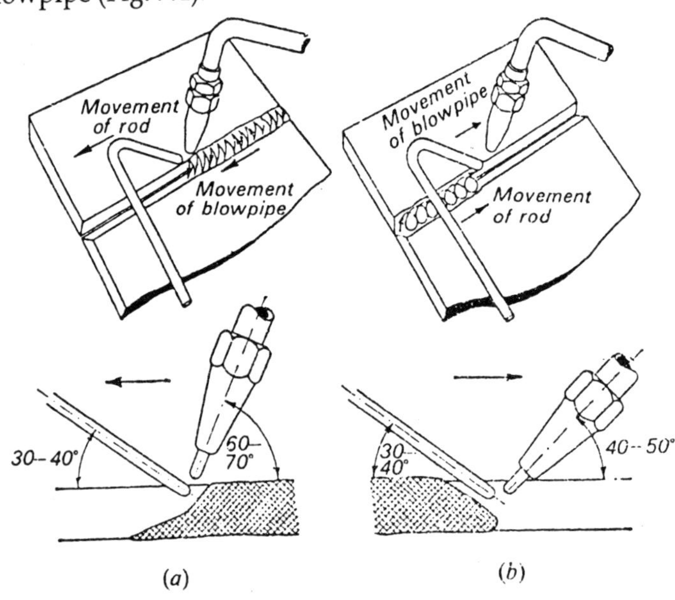

Fig. 7.4: Methods of welding: (a) the leftward, and (b) the rightward method

The leftward method is found most advantageous on steel plates up to 6 mm thick and on non-ferrous metals, while for steel work over 6 mm the rightward method is used.

Preparation of the edges to be welded is essential before making the weld. This consists for the most part in straightening, cleanign and, on thicker plates, vee-ing the edges. For thin plates and short runs the chamfers to form the vee can be filed or hand ground, but where posible they should be machined on a shaper, planer or milling machine. Table 7.1 is showing the suitable forms for the edges on plate, together with particulars of filler rod and blowpipe tip. For welding low-carbon steel the filler rod is approximately of the same composition as the work and is copper plated to avoid rusting.

Table 71.: Edge preparation, technique and speed–underhand butt welds in steel

Thickness of metal	Diameter of welding rod	Edge preparation
Less than 1 mm	1·5 mm	
1–3 mm	1·5–3 mm	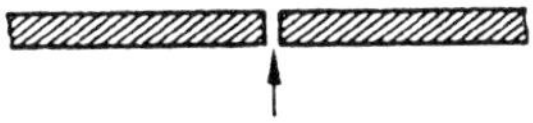 0·8–3 mm
3–5 mm	3–4 mm	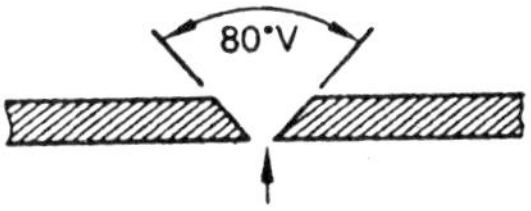 80°V 1·6–3 mm
5–8 mm	3–4 mm	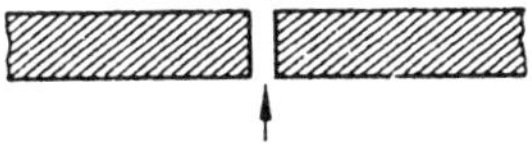 3–4 mm
8–15 mm	3–6·5 mm	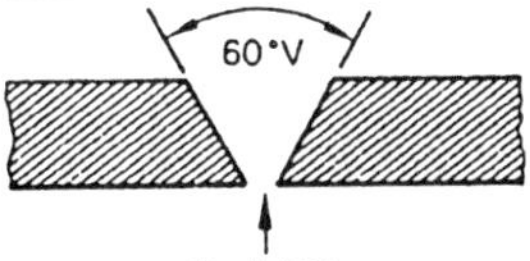 60°V 3–4 mm
15 mm and over	6·5 mm	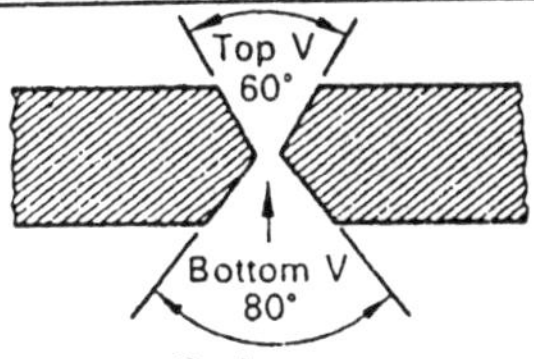 Top V 60° Bottom V 80° 3–4 mm

	Speed mm/min	Thickness of metal
Leftward welding	127–152	0·8 mm
	100–127	1·5 mm
	100–127	2·5 mm
	90–100	3 mm
	75–90	4 mm
Rightward welding	60–75	4·8 mm
	50–60	6·4 mm
	35–40	8 mm
	30–35	9·5 mm
	22–25	12·5 mm
	19–22	15 mm
	15–16	19 mm
	10–12	25 mm

British Oxygen Co. Ltd (some figures rounded off)

The desired manipulation of the blowpipe and filler rod and the various techniques require for making a good weld cannot be learned for the theortical knowledge of a text book. But these can be learned or acquired by the specialised practice under a good welder, so facilities of practice should be possess by a practitioner, but there are several important points which we might impress upon him. Since welding is essentially a casting process, care must be taken that the joint is made by thoroughly fusing the metals together and not by 'sticking' melted filler rod onto the surfaces to be joined. This will be avoied if care is taken to ensure that complete fusion penetrates the entire surfaces to be welded. If the material being joined is veed for the purpose of adding additional metal from the filler rod, the aim should be to concentrate the heat of the flame into the vee until a molten pool of metal appears, and then add metal from the rod by dipping the end into the pool so that it melts and makes up the vee. With practice this can be performed as a progressive movement along the weld. For thick plates it may be necessary to fill in the bottom of the vee with one pass and then repeat the process once or twice with additional layers of filling until the weld is completed. On no account should the work be hurried, but at the same time, particularly on thin plates, the flame should not be held too long in one place or a hole will be burned in the plate. As is shown in Table 7.1, thin plates are joined by melting their edges together without the addition of filling metal.

Fillet welding. Many details of construction entail the connection of two plates in such a manner that a weld in the corner is the easiest way of making the joint. In the under given

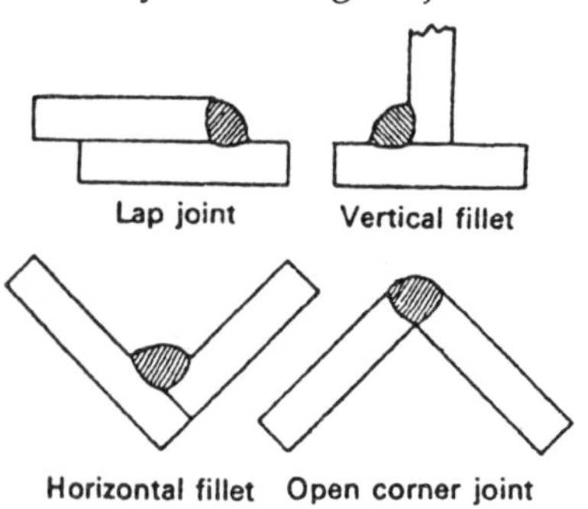

Fig. 7.4 Fillet-welded joints

Fig. the example of this work is shown. Usually the plates are first positioned and 'tacked' or slightly fused together at suitable points to hold them in position. The weld is then carried out by building up the fillet with additional filler rod. In making welds of this type, the precautions should be observed regarding fusion and penetration, but in addition the plates should not be unduly undercut by allowing the flame to dwell too long and melt away a large amount of the thickness.

Welding of cast-iron. Cast-iron welding can be performed satisfactorily, but there are certain prablems which make it difficult than the welding of steel. There are two chief difficulties given under.

(*a*) Iron never attains fluidity under the blowpipe in the same way as steel, but must be manipulated when in a pasty state. This makes fusion and penetration more difficult and promotes oxidation in the weld, as oxide cannot readily float to the surface of a pasty mass. For this reason a flux is necessary when welding cast iron. The welding flame also tends to burn the silicon out of the weld, allowing the carbon to assume the combined (carbide) form and render the weld metal hard when cold. This is overcome by using a welding rod rich in silicon so that the loss is replaced from the rod.

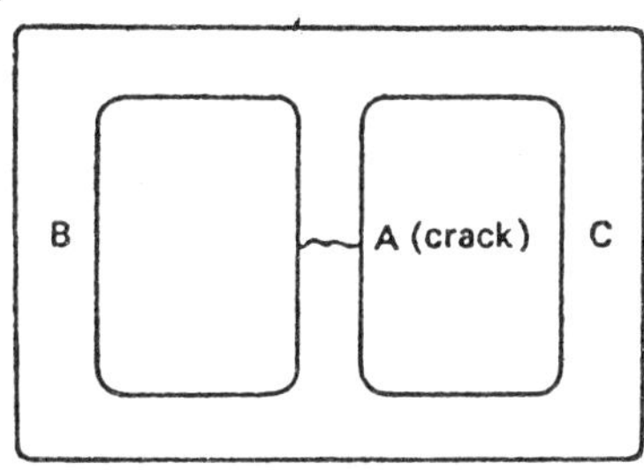

Fig.7.6: Cracked casting for welding

(*b*) Cast iron being a brittle material, the uneven expansion and contraction caused by local heating often results in racks occurring. This fault is accentuated by the fact that most cast-iron welding is for the repair of castings which may be of complicated shape. A simple example of this is shown in Fig. 7.6, which represents a casting to be welded at A. If this were done with the limbs B and C cold, when A cooled and contracted, a crack would occur somewhere in A (probably the weld) due to the rigidity of the remainder of the casting preventing unrestrained contraction in the centre limbs. The fault could have been prevented by preheating limbs B and C to a red heat, making the weld, and then allowing the three limbs to cool and contract together. A better method still would be to heat the whole casting to a red heat, make the weld, and then allow the whole to cool down slowly. This preheating is important on castings, and, if a muffle furnace or forge cannot be used, a temporary structure of firebricks can be built round the casting and heated with gas burners or a charcoal fire.

While we weld cast-iron, the practice of preparing the joint to the form of a bee is followed in the same manner as for steel. A bee should be cut on one or both sides at the crack according to circumstances. If the crack has to be made good, cutting should be followed by using file, chisel machine or other conventional methods. The poor fluidity and need for fluxing during welding render the operation less straightforward than for steel. Special cast-iron rod, rich in silicon, should be used and when pastiness has been obtained under the flame, the rod, previouslyy dipped in flux, is lightly stirred in the molten metal. This promotes fluxing and distributes the silicon being added from the rod. The slag and oxide on the surface of the finished weld may be removed by scraping, and brushing with a wire brush.

Malleable cast iron cannot satisfactorily be welded with cast-iron filling rod, but may be joined with bronze rod. Bronze welding may also be applied to cast iron and steel. In the cse of cast iron, the lower temperature necessary often dispenses with the necessity for preheating unless the casting is of complicated form.

ARC WELDING

A fusion process in which the welding rod is cast into the preriously fused space between the metals to be joined is known as arc welding. The heat necessary to melt the metal and welding rod is obtained from an electric arc struck between the rod (electrode) and the work, a temperature of 3500-4000°C being obtained near the crater of the arc.

Equipment for Arcwelding

Functions by are welding operatesd with diect or alternating electric current. In the direct-current (d.c.) welding sets the current is generated by a special generator with may be driven by an electic motor by some form of engine. Electrically driven genertors are ususlly employed in cases where the set can be permanently installed, and where ar supply of current is available. Engine-driven sets are most commonly driven by a petrol or diesel engine and may be of the permanently fixed type or may be mounted on wheels (or a motor chassis) so that they may be moved from place to place.

Alternating-current (a. c.) welding sets derive their current from an electrical transformer which takes in current at mains voltage (often 400 volts, and transforms it to the voltage necessary for he arc (80-100 volts). The cheaper and simple source of energy is considesed transformer there than generator set, because transformer set has not any movable part in it so, depreciation and maintenance is generally nil. These advantages are to some extent offset by the superiority of d.c. sets for welding cast iron and non-ferrous metals, and the necessity of always using covered electrodes when working with alternating current. The purpose of the regulators mounted on the top face is to provide the welder with a means for varying the electric current of suit the work in hand.

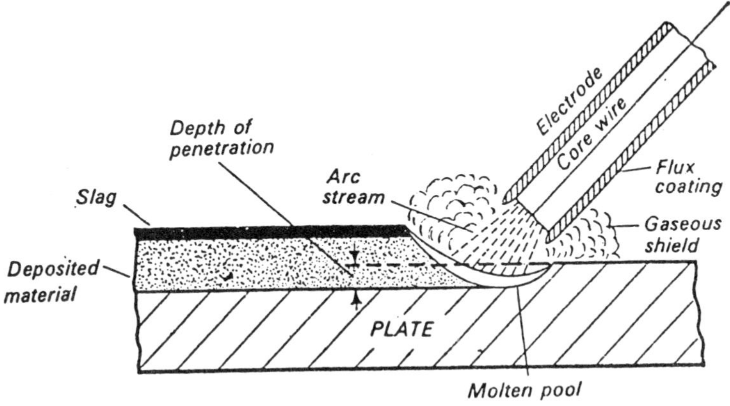

Fig .7.7: Conditions in arc welding with coated electrode.

The electrods are made in various lengths from 200mm to

450mm and in diameters form 3 mm upwards. Accrding to the work for which they are required, the composition is varies, mild steel, cast-iron, bronze and other composition may be there for a particular work.

Electrodes are nearly always coated with chemicals or fluex which enable a steadier arc to be maintained and, by fluxxes and covering the weld, promote improved quality in the weld. Electrodes may be lightly, medium or heavily coated. The lightcoated rods have a thin layer of a chemical such as lime which helps to stabilise the arc but has little effect on fluxing or slag formation. Medium and heavily coated rods have a covering of paste containing siliceous matter which melts and fluxes the weld, at the same time depositing a heavy layer of slag over the finished weld, reventing oxidation and reducing the rate of cooling. The covering also consists of substances which gic off gases and envelop the molten seat of the weld, preventing atmospheric oxidation while the operation is in progress. These rods assist and shield the are because the covering melts at a slower rate than the metal core, with the result that the casing projects beyond the molten end of the electrode metal (Fig. 7.7).

The electrode-holder gives object for clamping and holding the rod in a desirable position and has the necessary electricla connection to the set. The luminous intensity, of both the arc and the oxyacetylene flame, makes suitable protection for the eyes necessary. With the arc, an additional danger is the ultra-violet radiation which comes from it, necessitating a shield for the face and gloves on the hands.

Arc-welding practice

For welding purpose the preparation of the plates is very similar to the method for oxy-acetylene work, the thicker plates being veed on both side and the welding metal being deposited in a serties of layers (Fig. 7.8).

The electric potential require to maintain the arc dpends on the size and covering of the electrode. For bare of thinly coated rods the voltage varies from approximately 15 volts on the small to 20 volts on the large-diameter rods.

The figure are 20-25 volts and 25-40 volts respectively. For the medium and thickly coated rods to completing the electric circuit with one pole connected to the electrode, the other lead must be attached to the work. With d. c. workding, instructions as whether the electrode forms the positive or negative pole is usually given with the electrodes; with a.c. operation the polarity is immerial since with alternating current the polarity reverses at the speed of the frequency.

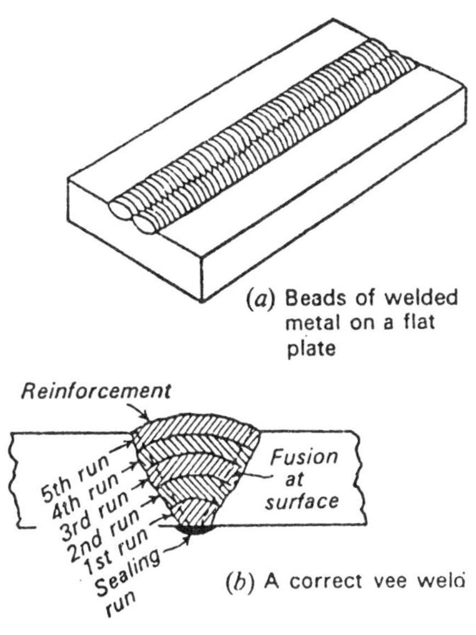

Fig. 7.8: Depositing weld metal

When comencing a weld the electrode, fixed in the holder, is held in the right hand and the face shield, if a hand pattern, is held in the left. The arc may be struck by jabbing the rod onto the plate and then withdrawing it to about 3mm to 6mm away, or by scraping the plate with it, using a slight circular motion so that the arc is struck at the bottom of the travel and further motion of the rod elongates the arc to the required lenght. When the arc is established, the rod is held at a fairly steep angle and then moved slowly and evenly towards the operator. The deposited bead of metal should be continuous and even, with good penetration into the parent metal. Fig. 7.8 shows beals on a flat plate and the correct method of filling in a vee.

Fillet and lap welding may be carried out with the arc method in the same way as for oxy-acetylene welding, and is shown in Fig.7.5.

Cast iron. While we weld-iron by the arc method. The electrodes used are either of a steel base with suitable alloying elements (e.g. silicon) or non-ferrous, such as phosphor bronze. The general method for the preparation of the weld is similar to that for oxy-acetylene work except that with arc welding no preheating is necessary unless the casting is very complicated in shape. This is because the heat is much more highly concentrated and localised in arc welding, a factor which often gives the arc method a great advantage over oxy-acetylene working. Often a weld has to be made near to a bearing or to a finished surface which would be spoiled by undue heating. Oxy-acetylene welding would, in all probability, result in such a spread of heat that the work would be ruined, but by using the arc method metal at only a small distance from the weld will be left comparatively cold.

Building up and surfacing

Oxy-acetylene method and electric method both are used to deposit layers of metal on lop of an existing surface. These techniquers has a great importance not onl in repair works but also in many production works. Welding enables the worn end of a shaft to be built up with layers of welded metal so that it can be turned and ground to its original size.

Stelliting–An Allloy of Cobalt

Anything which requirer a very hard and wear-resisting surface may have a layer of stellite deposited by the oxyacetylene or by the arc method. Such as cutting tools, rock dulls, press-tool punches and dies and knife edges, etc. Stellite is an alloy of cobalt, chromium, tungsten and carbon which lends itself to the casting process and is procurable as wlding rod. After the surface has been deposited, the stellite may be ground to the finished shape required and will give a surface having many times the life of hardened steel. Fig.7.9 shows a blanking die having its edges made up with stellite.

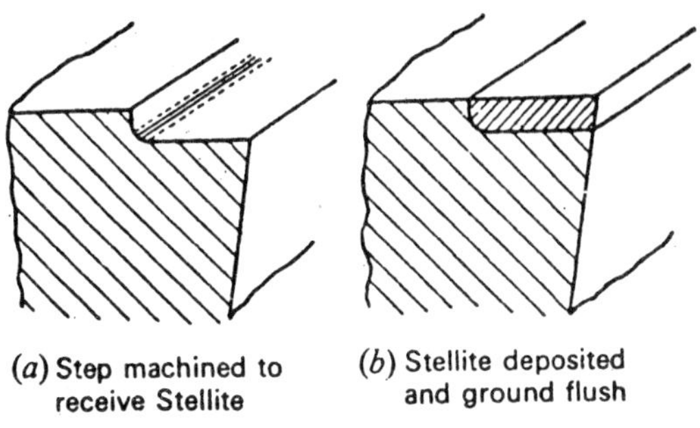

Fig. 7.9: Stelliting edge of die

SPECIAL ARC-WELDING PROCESSES

We have discuss earlier the metal arc metod, it is most common welding process in general engineering. But on other hand it has so many disadvantages, and, to widen the scope and efficiency of welding, various special processes have been developed. These, in the main, are directed towards adapting the metal-arc process to mechanised, automatic operation and to overcoming the difficulties associated with certain metals.

Carbon-arc welding

Between the carbon electrode and the parent metal an arc is struck. The generated heat is melts some of the parent metal and also, where necessary, a metal filler rod. The gas envelope from the burning carbon serves to some extent to protect the weld from the atmosphere, but it is usual to employ a flux to counter-act carbon pick-up from the electrode. This can also be minimised by using d.c. welding, with the carbon electrode connected to the negative pole of the generator. When used manually, this method does not offer any particular advantages over metal-arc welding and its chief possibility lies in its development for automatic operation. When used in this way, a mechanism is necesary for holding and feeding the carbon rod, the flux and the filler rod. This last can be in the form of a coil mounted on a drum. Automatic carbon-arc welding is capable of high-speed working and has been used for mild and low-alloy steels, copper copper and some alloys, nickel and Monel.

Argon-arc welding

In argon-arc welding an arc is struck between a tungsten electrode and the work, the electrode bing mounted in a nozzle-shaped hood through which the inert gas argon is passed at a low speed (0.2-0.7 bar pressure). The shield of argon gas completely protects the weld from the action of the atomsphere and no fluxin is necessary. The electric power supply can be from a d.c. generator or from an a.c. transformer. If a.c. is used, auxiliary equipment is necessary to maintain the arc during the periods that the electric current passes through its zero. This

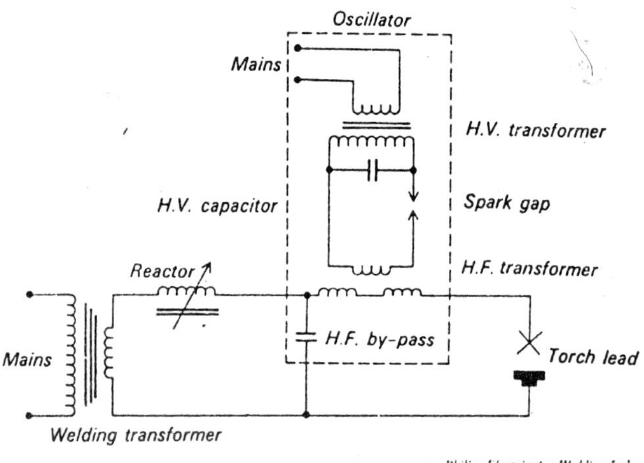

Fig. 7.10: High-frequency injector unit

consists of a high-frequency, low-current sascillator which is connected into the system as showing in the obove given figure. Although this unit is not necesaery with d.c. working, it is often used as it facilitates the striking of the arc, a more difficult undertaking with bare tungsten than under normal welding conditions. Two sizes of torch are available for hand welding, having capacites of 150 amperes and 300 amperes, the later being water-cooled. For automatic working on long runs a 600 ampere water-cooled torch is available. The tungsten electrode, which has a melting point of 3350°C, does not melt in the arc, although a slight loss occurs in the course of time. The manual operation of argon-arc welding is conducted in much the same way as we have discussed for metal-arc working, where the torch is held in the right and the filler in the left hand.

An important favourable point of this process is the freedom from the necessity of using a flux. Many of the most effective fluxes are corrosive and this precludes their use for many purpose, e.g. in the food, drink and certain chemical industries. In aluminium, magnesium, copper and other non-ferrous metals, this process provide clean and sound welding.

Atomic-hydrogen welding

The process of atomic hydrogen welding is depends on the on the dissociation of hydrogan electric arc and when the atomic gas recombines to its molecular state, the enormous heat is released. A slow stream of hydrogen is passed through an arc struck between two tungsten electrodes. The arc dissociates the normal molecular hydrogen (H_2) into atomic hydrogen (H_1) which passes on to a cooler region just beyond the arc, where it recombines into its former state. This liberates heat to produce a temperature of the order of 3700°C and, furthermore, the high conductivity of heat by hydrogen in this state gives a very high rate heat delivery (about twice that of oxy-acetylene) to the work. The equipment for the process consists of the electrical plant for producing the arc, the gas plant for a supply of hydrogen, and the welding torch. Normally an a.c. electric supply is used, since better current regulation is possible, and the reversing polarity equalises the electrode consumption between the two tungsten rods.

Submerged-arc welding

The process of submerged arc welding has developed as a vesult of the need to produce high-quality welds by an automatic mechanised process under the best possible situations. The arc between the welding rod and the work travels beneath the surface of a heap of flux which is fed from a tube mounted on the welding head and travelling in front of the electrode wire. This flux is a granulated powder form of calcium/magnesium silicate which fuses at a temperature below that of the weld. This results in the weld being submerged in a bath of molten slag with a thick layer of unfused flux on top, resulting in a high localisation of the heat and complete protection from the atmosphere. As the weld proceeds, the fused flux solidifies to form a protective covering slag which carcks off on cooling, the unfused material being collected and used again. The welding rod is in the form of copper-plated steel wire carried on a drum and fed through rollers. The process is completely devoid of flash, spatter, smoke and noise, and there are hardly any fumes. It produces high-quality welds and so far has been used mostly for carbon and alloy steels, although its application is likely to spread to other metals also. Normally applied to automatic operation, it may be operated by a travelling head on long flat runs or with a fixed head and travelling work for welding pressure vessels. Equipments is use generally work from an a.c. supply and welding currents up to 3500 amperes are possible. No other method is capable of such high speeds and heavy currents, and if desired a thickness of 50 mm can be welded in one pass. A diagram of the process is shown in Fig. 7.11.

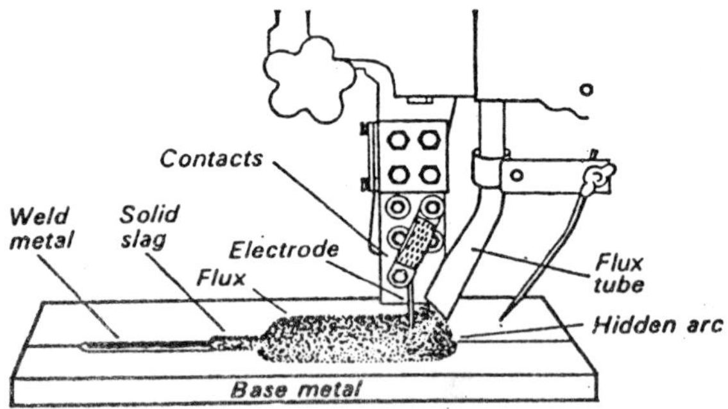

Fig. 7.11: Diagram of the submerged-arc weld process

Metal inert gas (MIG) welding

In the metal inert gas (MIG) welding process, the coalescence is obtained by striking an arc between the workpiece and a continuous wire electrode fed through a torch. A shielding gas flows through the torch and forms a blanket over the welding puddle to shield it from atmospheric contamination. Following the certain equipments which are require for MIG welding purposes:

(i) *Power source.* A transformer-rectifier producing d.c. variable-voltage output and an auxiliary power supply for the wire-feed motor and gas heater.

(ii) *Feeed unit.* This incorporates the wire-feed motor and wire spool and monitors a constant wire feed to the torch nozzle.

(iii) *Torch.* This is connected to the end of a flexible conduit carrying the welding current, the electrode wire, and the shielding gas.

(iv) *Shielding gas supply.* This may be carbon dioxide, argon, or argon-rich mixtures, according to the work. The gas is normally used from standard cylinders with regulators. This has a capability up to 33 amperes and can be used on all weldable materials from 1 mm up to any thickeness.

In MIG working, there are two basic modes of metal transfer. In *dip-transfer,* the electrode wire dips into the weld pool, causing a short-circuit of sufficient duration to extinguish the arc. This results in a rapid enhancement in the current which heats up the electrode wire causing it to melt off and re-establish the arc. This is repeated apporoximately 100 times per second. The weld pool is small, fairly stiff, and therefore ideal for all positional welding of sheet metals.

The spray-transfer method may be used, for dip the transfer. In this situation metal droplets are transferred into the weld pool without any short-circuitin. The pool is relatively fluid and the process is ideally suited for fast welding on thicker sections. The outstanding feature of metal inert gas welding is the production of high quality welds at high welding speeds without the need of a flux or the necessity for any slag removal. For these reasons, and for its versatility, it has superceded alternative welding methods and is widely employed in almost every industry. By this process we can weld most metals, aluminium, carbon and low-alloy steels, nickel, copper, magnesium, titanium, and zirconium.

Stud welding

The stud is a very common object for student of engineering, and the time taken to drill and tap the hole to accommodate it is considerable. Furthermore, when a stud has to be attached to a thin material the length of thread engagement may by not sufficient. The attachment of studs by welding is carried out with a gun which holds and positions the stud. Controls on the gun switch on the current and withdraw the end of the stud sufficiently for the arc to be struck, and, after an arcing period sufficient to form a pool of metal, a spring presses the end of the stud into the pool. The total time taken to fix a stud is less than a minute and studs up to 25 mm in steel and 10 mm in brass may be welded. Location may be effected by marking out and centre punching, but for any degree of accuracy it is desirable to use a plate jig. Special studs are used for the process, sometimes being coated with aluminium, which is beneficial to the arc. The process of stud welding can be used for carbon and alloy steels brass, bronze and aluminium alloys. To avoid stress concentrations it is advisable not to weld studs threaded their full lenght, but to have a plain, unthreaded length of the stud adjacent to the weld.

Table 7.2 : Stud welding–steel studs

Stud diameter (mm)	10	13	16	20
Average welding current (amperes)	750	1000	1500	1800
Arcing time (seconds)	0.3	0.4	0.5	0.6

Resistance welding

In this type of welding the two electordes or two clamps are incorporated in a circuit of low resistance and the welded metals are pressed between the electrodes and pressed together. The circuit is thus completed and the electrical resistance at the joint of the metals to be welded is so high, in comparison with the rest of the circuit, that if the curent is heavy enough the joint of the metals is heated to the plastic state necessary for welding to take place. The welding ciruit usually form the secondary circuit of a transformer, the primary of which is fed with alternatihng current from the mains, as shown in Fig. 7.13, which shows the arrangement for spot welding.

Fig. 7.12: Stud weld

Two specific windings that are primary and secondary, are lacated on an iron core, the function of which is to enhance the magnetic effect upon which the induction of current depends. One of these windings has a larger number of turns of wire than the other, the ratio depending upon the input and output voltage ratio required. If an alternating voltage is applied to one of these windings, whern the circuit containing the other one is closed a current will flow in it and the voltage will be related to the voltage applied to the other winding in the inverse ratio of the turns.

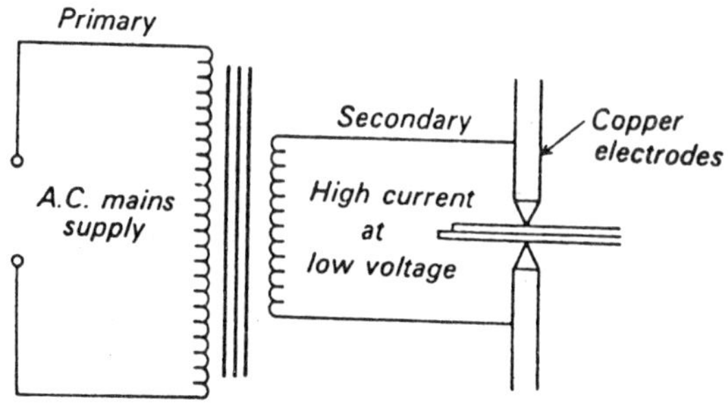

Fig. 7.13: Electric circuit for spot welding

The welding transfer is shown graphicallly below. In this scatch you would learn that the primary winding is fed from the mains at 400V, and the secondary forms the helding circuit in which a voltage of about 10 is required to complete the welding. This gives a ratio of 400/10=40 and the current will be stepped up in this ratio. Thus if a current of 50 A flows in the primary, the current in the secondary will be approximately 50×40=200A. Naturally, the wire in the smaller coil will be thicker than that in the other, so that the heavier current may be carried.

The main types of resistance-welding machines are spot, seam, projection and butt welders. In the spot welder a pair of copper electrodes, air- or pedal-operated, press together with the metal to be welded in between, and a local 'spot' is welded at the point where the metals are pressed by the electrodes, the effect being similar to a rivet. In the larger types, the electrodes are water-jacketed to keep them cool under the heating effect of the heavy current flowing.

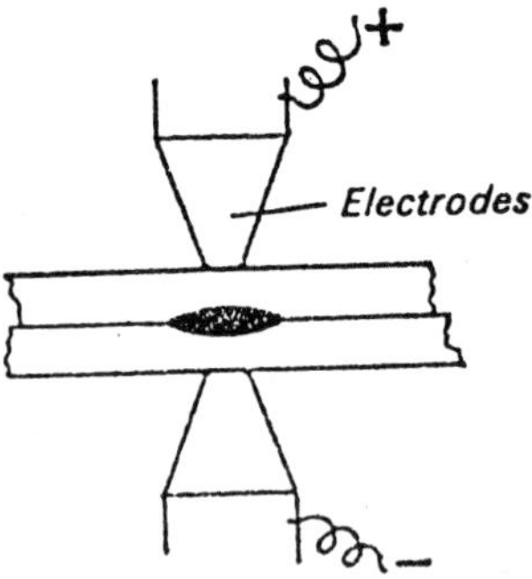

Fig. 7.14: Pedal-operated spot welder (45 kVA maximum rating). Water tube connections can be seen on ends of electrodes.

Except that instead of two pointed etectrodes there are two edged rollers in the seam weldes. The plates to be joined are passed between these and, while a low motion takes place, the current is switched on and off at a suitable rate by some means such as an interrupter switch. This results in a line of spot welds, the spacing of which is determined by the speed of movement and the current timing.

The projection welder enables a number of welds to be made simultanously, often by automatic indexing and control. This process is admirably suited for certain applications of mass production.

The flash butt welder is a totally automatic machine with pneumatically operated clamps of the two piece parts, a rotary motorised cam for flashing the workpieces to fusion temperature and an air cylinder for supplying the upsetting force immediately after flashing. It is useful for welding a high-speed-steel body to a cheaper carbon-steel shank in the manufacture of large twist drills, for welding high-speed ends

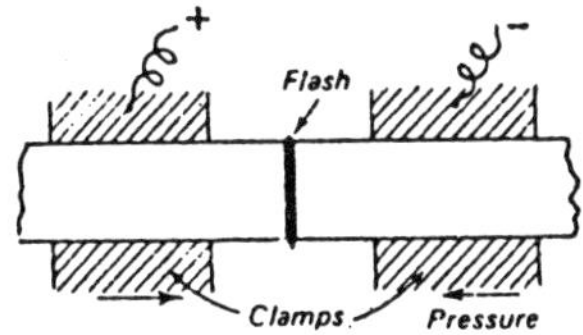

Fig. 7.15: Flash butt welder (150 kVA, max. welding capacity 9.5 cm^2 mild steel) and diagram of butt welding

to carbon-steel shanks on turning tools, and for other applications of a similar nature.

These days in the process of mass production the resistance welding methods have been applicable extensively. This become possible after developement of the machines for automatic and semi-automatic operations or functions. Many fastenings for which rivets and other methods were formerly used are now made by resistance welding, owing to its cheapness and speed of operation. Examples are too numerous to mention, but the reader may see many of them for himself if he studies the construction of mass-produced sheet-steel and wire articles.

OXY-ACETYLENSE CUTTING

Basically iron and steel may be cut with a special blowpipe, and the process provides a means of cutting which is fast, clean and capable of following almost any profile. In this last respect

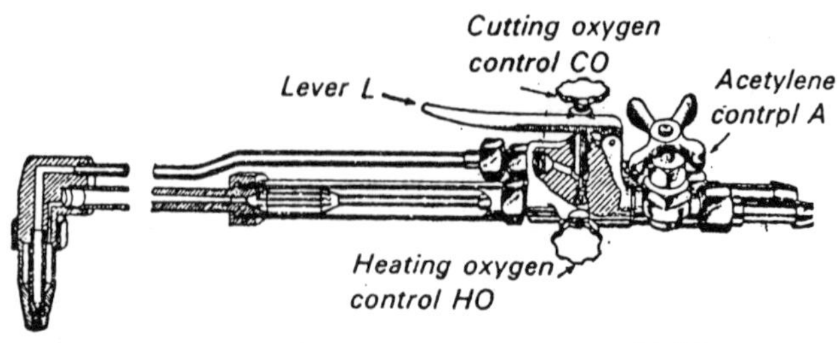

Fig. 7.16: Oxy-acetylene cutting blowpipe

it has a great advantage over the usual cutting methods, most of which lend themselves only to making a straight cut perpendicular to the length of the material. A diagram given on the ahead page is represented the scatch of the cutting blow pipe. In the Figure it is seen that there are two saparate gas passages: the central bore for cutting oxygen, controlled by the hand lever at the end of the blowpipe, and the outer annular space for the oxy-acetylene heating-flame gas, controlled by the usual valves.

When we operate the torch, the heating flame is lit and adjusted until there is a short blue luminous cone, and the guide is set so that when it is rested on the plate the nozzle is about 6 mm from its surface. Staring at the edge of the plate or bar to be cut, the metal is brought up to a white heat with the heating flame and then the valve controlling the cutting oxygen is released. At the same time, a steady feed of the blowpipe towards the operator is begun and the cut will proceed. A guide, clamped to the nozzle, serves as a support to maintain a uniform distance from the nozzle to the plate and may also be used to follow the marking out of any special profile.

The function of the cutting blowpipe is that the jet of high-pressure oxygen being directed onto the metal oxidises it, and because the oxide has a lower melting point than the surrounding steel, it melts quickly and is blown through by the stream of oxygen. The edges of the cut will be subject to some hardening, due to the rapid rate at which the heat is conducted to the surrounding metal. This effect will depend on the carbon content of the seteel, but on steels below 0.3% C it is not likely to be troublesome and only penetrate about 3 mm.

The cutting blowpipe has been adapted to special cutting machines for producing profiles in steel plate. Generally the arrangement is such that the blowpipe is connected through a pantograph mechanism to a tracing pointer which, when moved round a template, causes the blowpipe to follow its path and reproduce the profile as a cut in the metal plate.

HORIZONTAL FORGING AND HEADING

Forging, upsetting and heading machines which operate on a horizontal, crank-driven principle and which produce forged shapes either on or from the bar constitue an important group of machines for the preparation of primary forgings. When a component such as a headed bolt or a gear on the end of and integral with a shaft is formed by upsetting from the original shaft size, a result is obtained much supperior to that when the article is turned from the solid. Not only does the process of forging improve the mechanical properties of the material but also the direction of grain flow promoted by the process results in improved strength and directional properties. This is shown by the macro-etchings which indicate the grain flow that has been induced by the forging process and which is superior to a grain directed in straight lines parallel to the centre line as would be the case for the same articles turned from bar. Some forgings of this class, such as small gears, ball-race housings, hubs and other cylindrical shapes, are forged by upsetting the end of a bar and are then severed from the bar by punching off. In such articles we may expect the same beneficial qualities as we have discussed. An additional factor not to be ignored is the economy in material offered by the process. In general, all the material used goes into the forging, and , since this is produced in a closed die, very close limits of size, and high quality of surface are possible. These factors operate in favour of the minimum of machining, and of reduced allowances where this is necessary.

Horizontal forging machine

The main bed frame is a heavy box-shaped casting with a tie bar on the larger sizes to assist in carrying the forging pressure. The heading-slide casting bridges over the main crankshaft to provide an additional bearing slide, as seen to the extreme right. The die, shown to the left of the punches, is in two halves, divided down the centre of the work impressions. One part of the die is fixed to the frame and the other to a slide so that the halves may be opened to insert the bar.

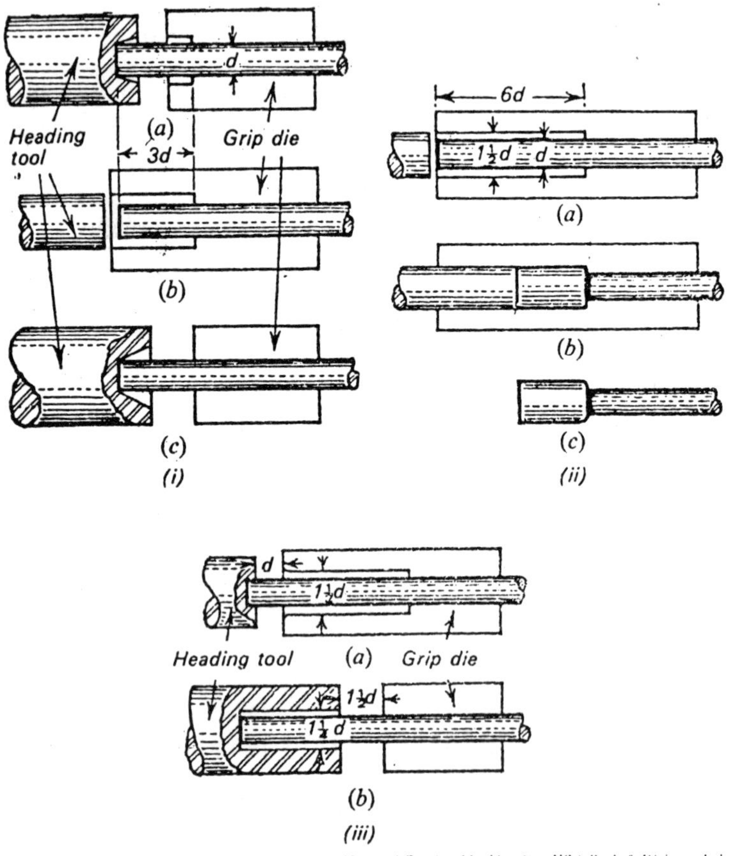

Fig. 7.17: To show the rules for upsetting

Forging Practice

The end of a bar is heated to forging temperature for the required length, in the preparation of a forging. It is then inserted between the (open) dies, located against the stop, and positioned so that when the dies close it will be taken in the impression for the first operation. Operation of the foot pedal starts the machine, the dies close on the bar and the punches advance until they have completed their travel, when the first operation punch has completed its work. The first cycle is completed by the return of the heading slide and the opening of the dies. The work is then transferred to the second operation station and the cycle is repeated for this position, and so on.

The work of these machines is highly specialised, and in the space at our disposal we are able to indicate only the bare outline of the process. The successful design of the process is governed by the three sets of conditions (see Fig. 7.17).

1. Fig. 7.17 (i). The maximum length of stock that may be gathered or upset in one blow without buckling is three times the diameter of the bar.
2. Fig. 7.17 (ii). Lengths of bar greater than three diameters can be upset successfully to diameter only 1½ times that of bar.
3. Fig. 7.17 (iii). In an upset length of more than three bar diameters, the diameter and unsupported-length relationships are as shown in Fig. 7.17

Operational Planning

The above rules are followed in plainning the operations and designing the tools in respect of the particular job concerned.

Forging Tools' Materials

The dies and punches used for the hot working of metals are subject to extremely arduous conditions of service, and the selection of suitable steels is of great importance. The three chief causes of tool deterioration are *pressure, abrasion* and *heat,* and, since the tools are in contact with red-hot material for considerable periods, a steel which is able to with stand pressure and abrasion in the cold state might be totally unfit under this additional condition.

The extent of the deterioration depends on various factors involving conditions and designs. For example, a shape having a large area in proportion to its volume, or a component whose volume is small compared with that of the dies, permits heat be dissipated more quickly and conserves the dies. Die impressions with thin or sharp edges are bad, since the small, local projections are liable to burning, loss of temper, and some decarburisation.

Plain-carbon steel is suitable only where the heating effect is a minimum, and for more arduous conditions alloy heat-resisting steels must be employed. Tungsten and chromium are the two alloying elements found to be most-effective for the purpose. The tungsten steels used are of the high-speed variety with about 0.4 to 0.5% carbon, up to 18% tungsten, and 3 or 4% chromium. A chrome steel without tungsten may contain 0.8 to 0.9% carbon, 3 to 4% chromium, and 0.5 to 0.6% manganese. Heat treatment of the tools according to the steel-makers' instructions is necessary. A set of tools for forging a ring and the lettering shows the class of steel recommended for each part of the tools.

Cold heading and forging

Cold heading and forging is a process performed on a heading machine, the general design of which is very similar to the hot forging and extruding is possible, machine we have discussed. As its name implies, however, the forging is performed with the metal in its normal, cold condition; so the process is confined to smaller bar sizes and simpler upsets than with hot forging. Cold-heading machines are constructed in sizes up to that capable of producing articles from 20 mm bar and upwards, and most of them are designed for automatic operation. For material in the form of a coil, this is carried on a drum and is fed to the tools by feed rolls, geared together and actuated by a pawl and ratchet wheel or by some other suitable method.

Alternative feed arrangements are used for larger material in bar form. Since cold headers work automatically, their operation differs in several respects from hot-forging machines.

(*a*) The die is provided with only a single work station and is often made solid (i.e. split down the centre). In addition to its working bore, however, the die carries another hole through which the incoming material is fed.

(*b*) Cut-off and transfer mechanisms are provided to shear off a suitable length of material at the face of the feeding station, and to transfer the blank, so cut, to the working line of the punch and die.

(*c*) When the header slide carries more than one punch, these are not fixed in relation to corresponding die stations but are carried on a vertical slide with a mechanism provided to bring each punch, in turn, opposite the die.

(*d*) The finished part is backed up in and ejected from the die by a cam-operated ejector. The arm near the wire is the shear blade. In operation the lower, V-shaped arm rises under cam actuation and the wire is gripped between this and the shear blade.The two arms gripping the wire then swing down, shearing off the blank and carrying it to a position opposits the die where it is picked up by the punch and conveyed into the die, the two arms meanwhile moving clear.

Heading machines are widely dused for the production of bolts, rivets, headed pins and similar work where the component consists of a shank bearing some form of enlarged head.

Other than a certain amount of swaging and extruding is possible, as is necessary for a chain rivet with slightly reduced ends or a bolt with a reduced portion for a rolled thread. The question as to the choice of solid or open die; one-two or three-blow operation; etc. will depend upon the material being worked and the form of the upset to be obtained. Hard, alloy steels will necessitate more blows than softer material for a similar result. Solid die machines are used where the component is not more than 8-10 diameters long, and open dies for greater lengths. The number of blows is determined by the length to be upset, and a general rule is one blow is determined by the length to be upset, and a general rule is one blow up 2½ diemeters, two blows for 2 ½ -4½ diameters, and three blows for 4½-6 diameters.

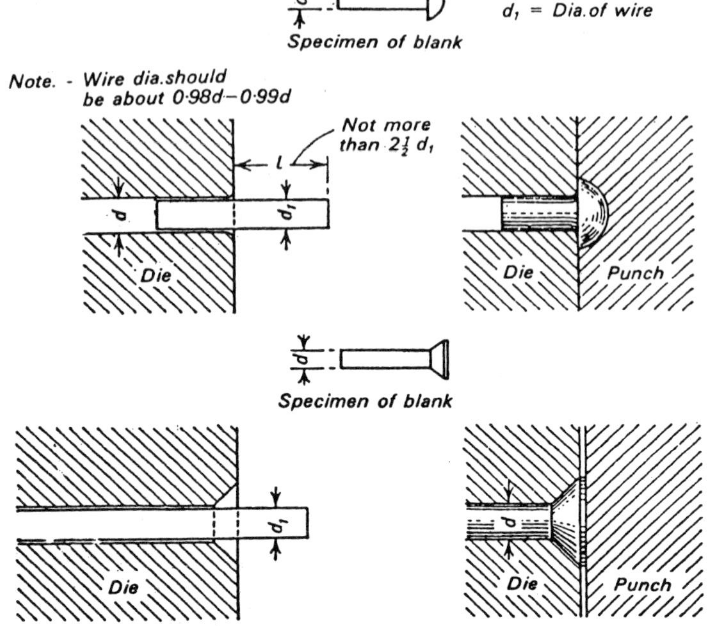

Fig. 7.18: Tools used on single-blow cold header

Some Examples of work

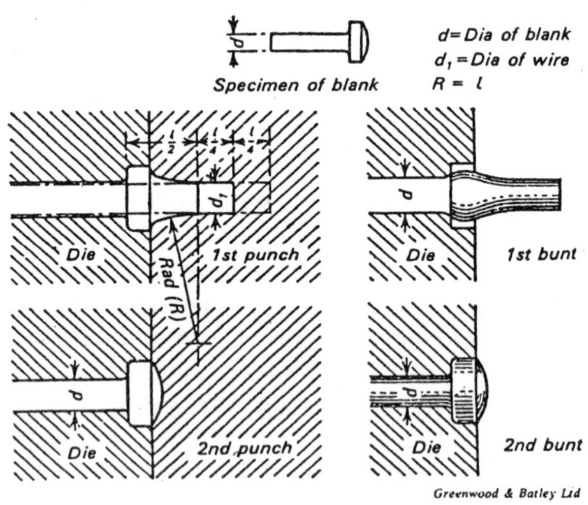

Fig. 7.19: Tools for two-blow cold header (heading die recessed)

Some of the examples of single-blow work are shown in Fig.7.18. Normally the wire is slightly less than the finished shank diameter, allowing for some compression and enlargement to take place in the die.

See Fig. 7.19 in this figure the two-blow layout for producing a cheese head when the head is formed in the die.

PRECISION INVESTMENT CASTING

The materials development and accuracy accentuated the need for casting methods superior to the traditional practice of sand moulding. Investment casting is not new and for many years has been common practice in the jewellery and dental industries. Its application to engineering has led to the development and adaptaion of various processes to cope with the different condition involved. The term 'investment' refers to the layer of refractory material with which the pattern is covered to form the mould. This form of casting has one element in common with ordinary sand casting in that the mould in destroyed each time a casting to mode. Die casting, as we shall discuss later, employs a permanent metal mould.

Shell Moulding

In the process of shell moulding, the pattern and shell are then heated sufficiently to cure the shell into a rigid state. The shells, made in halves on a parting line, are then clamped together for pouring the metal of the casting. Patterns may be machined from copper alloys, cast iron, steel or aluminium, depending of the life required, wear, cost of machining, and so on. They are made in two halves, with shrinkage allowance and draft, and must be accurately matched and polished. The two half patterns are each attached to a metal pattern plate through which pass a number of ejector pins for stripping the mould off after curing. The sand used for investing the pattern should have a high silica content and be dry and free from organic matter, with the minimum of clay, metallic oxides and alkali. The size, shape and distribution of the grains are important, since these influence the mould strength and casting surface and the percentage of binder necessary. Plastics resins (thermosetting) are used as a binder, and various special powders are available which must be mixed thoroughly with the sand, in a ratio varying from 5% to 10% by weight.

This is a rectangular steel box, open at the top and mounted on trunnions so that it may be rotated to the upside-down position. The open top of the box is the same size and shape as the pattern plates and is provided with clamps for securing the plate and sealing arrangements to contain the sand mixture. The box is kept charged to about one-third full with the sand mixture.

To invest the pattern, this with its plate is heated to about 150 – 230 °C and, after spraying with a silicone parting solution to minimise adhesion of the mould, is clamped, pattern inwards, to the open end of the box. The box is then reversed on its trunnions so that the sand contents fall and bury the pattern to a depth of about 100 – 150 mm.

In less than a minute, depending on the temperature of the pattern, the plastics partially thermosets and the pattern will have acqurired a shell of the sand about 5 – 7 mm thick, which is the normal thickness for the mould.

The pattern and plate are then removed from the box and transferred to an oven at about 250-400°C to complete the curing of the mould. This process takes 1 – 5 minutes and has the effect of developing the property of the resin content of the sand mix of binding the mould into a rigid shell capable of withstanding the casting stresses. This completes the preparation of the mould and, after stripping the shell from the plate, the casting is made by clamping the two half moulds together, pouring the molten metal, and stripping the casting in the usual way.

The process lends itself to the use of cores, which are made in hot metal core-boxes using a similar sand mix. As the moulds consist of only a thin shell, cases arise with large castings when the shell may not have sufficient strength to withstand the surge and weight of the hot casting metal, and this is overcome by enclosing the mould by an open box and packing sand round it for support.

Machines are available for preparing shell moulds in an automatic cycle. Such a machine would incorporate the dump box, curing oven and means for heating the pattern plate, with the necessary mechanism and timing control to complete the cycle after the pattern plate has been inserted. Shell moulding produces castings of exceptionally good finish with a dimensional accuracy of 0.3-1% on castings ranging up to 100 kg in mass. It can be used to cast most metals, including cylinder iron and steel, and, owing to the high porosity of the mould, castings having very thisn sections are possible.

THE 'LOST-WAX' PROCESS

By way of this procese the refractory moulding prepared through the agency of a wax pattern which is melted out after the mould has been made. The production of a casting is effected in three stages: (1) making the wax pattern

(2) investing the pattern to make the mould and

(3) pouring and stripping the casting.

For the preparation of the wax pattern a split master die must be made, and this can be done either (*a*) by preparing a master pattern and using it to cast the die, or (*b*) by machining the master die with an impression which is a negative of the pattern.

In the components master pattern is first made in an easily machined metal such as brass, steel, aluminium, etc. with allowances for shrinkage and the changes,which may occur in the refractory mould during firing. This is then embedded to the parting line in plaster of Paris in a moulding flask and placed in a pressurised chamber, and the molten alloy for the die is poured until it covers the pattern by about 25 mm over its highest point. By reversing, and pouring over the opposite side of the pattern, with a parting compound applied at the joint, the two halves of the master die are produced. Suitable alloys for casting split-master dies include type metal, Wood's metal and certain Cerro-base alloys.

Methods for die sinking must be used. The dies can be made in metals harder and more durable than the soft casting types necessary for (*a*) above, and are therefore more permanent. On the other hand, of course, even if the soft-cast dies wear out more rapidly, the master pattern is available for casting replacements. The requirements of the expendable wax pattern made in the master dies are (*a*) sufficient strength to withstand handling during investment (*b*) it must melt completely from the finished mould without leaving residue or ash, and (*c*) it must have a low shrinkage. Two general types of material are used, (1) waxes and (2) plastics. Waxes used are blends of beeswax, paraffin, cersein, carnauba and resins, the chosen blend being suitable for the characteristics desired, e.g. castings having fine detail, large castings, etc.

Plastics employed for this work include polystyrene, methyl methacrylate and the polythenes, the first being most common since it can be injected into the dies with standard equipment. To make the patern, the wase is injected into the dies under presure often the die surfaces have been smeared with condensed ethyl silicate to prevent adhesion. For wax a pressure of about 4 bars is used, with the dies heated to about, 65 °C, while the pressure for plastics fillings is of the order of 35 bars with higher temperatures.

The investment of the wax pattern takes place in two stages. First the pattern is precoated by dipping or spraying with a slurry made by mixing a fine silica flour with a water/ethyl-silicate solution. This coating is about 1mm thick and serves as an interface between the delicate surface of the pattern and the coarser grains of the main investment.

Secondly, when the precoat is dry the investment is carried out by inverting the wax assembly onto a metal plate and surrounding it by a paper-lined flask or a cardboard tube. The whole is then transferred to a vibrating table, while the investment slurry is poured on from a height of about 150 mm. The investement is composed of silica or quartz grains and binders, variations being necessary according to the pouring temperature of the casting metal. After investement, the flask and plate are transferred to a drying oven at about 38 °C for a period up to 48 hours and, after that, with the mould still invested, heated to about 200 °C, at which temperature the wax melts and runs out, being salvaged for future use.

The final treatment of the mould is that of firing, which consists of raising the temperature in a muffle furnace slowly (100 °C per hour) up to 850 – 1000 °C. During this period any residue of wax is burnt out, and the investment is rendered to a hard, rigid state. After firing, the mould is not allowed to cool down, but is transferred direct for casting, which in this process often consistes of either a pressure-casting unit or centrifugal-casting equipment. Large, non-ferrous castings are sometimes poured direct from the ladle, but supplementary air pressure may be necessary to fill an intricate mould.

The high quality of the casting surface and the close limits possible have made the lost-wax process important for eliminating complicated and costly machining processes. In cases where intricate parts have to be made in one of the modern heat-resisting steels or nimonic alloys, which are almost impossible to machine at all, the process has provided a solution to what was an almost insoluble problem.

SOME OTHER PRECISION CASTING METHODS

Plaster moulds

Some of the metals like copper and aluminium alloys may be cast in moulds consisting of plaster of Paris, with ingredients such as asbestos, silica, etc. to give the required mould qualities. Metal patterns are generally used, and the mould is made with a parting line in a flask. The drying time (10-20 h at 200 °C) is rather lengthy, but because of the dimensional accuracy and surface finish obtainable the method is useful for certain problems.

The Mercast process

This process follows the general principle of the 'lost wax', in that the pattern is melted out after the mould has been made. In uses mercury as the pattern medium. This freezes solid at 40°C, so the process up to the stage of drying and firing must be carried out at a temperature somewhat below this. The use of frozen mercury gives superior smoothness, and accuracy both of detail and size.

Die casting

In permanent steel moulds (dies), die castings are produced, in contrast to end eastings where the mould is prepared in sand with the necessity of destroying and remaking the mould each time a casting is produced. The development of die casting has made an enormous contribution to production engineering, as the process gives castings having many qualities not attainable by sand casting.

Die castings, unlike sand castings, are, as a general rule in their finished state when they leave the die and require little or no subsequent machining. The speed with which they may be produced, and the consequent lowering in cost, has resulted in the process being applied to an extensive range of products, including such articles as toys, chassis for radio sets, camera bodies, engine cylinder heads, gears, domestic-appliance components, and so on.

Die-casting metals

Although die-casting has been applied to a limited extent to iron, the chief metals being cast at the present state of development are certain non ferrous groups, the two most common being the aluminium and the zinc-base varieties. Among these are alloys having the approximate compositions shown in Tables 7.3 and 7.4.

Aluminium alloys

Table 7.3: Aluminium die-casting alloys (BS 1490:1970)

BS designation	LM2	LM6	LM20	LM24
Percentage of				
copper	0.7-2.5	0.1	0.4	3.0-4.0
magnesium	0.3	0.10	0.2	0.1
silicon	9.0-11.5	10.0-13.0	10.0-13.0	7.5-9.5
iron	1.0	0.6	1.0	1.3
manganese	0.5	0.5	0.5	0.5
nickel	0.5	0.1	0.1	0.5
zinc	2.0	0.1	0.2	3.0
lead	0.3	0.1	0.1	0.3
tin	0.2	0.05	0.1	0.2
titanium	0.2	0.2	0.2	0.2
aluminium	remainder	remainder	remainder	remainder

Zinc-base alloys

Table 7.4: Zinc-base alloys (BS 1004:1972)

BS designation	Alloy A	Alloy B
Percentage of		
aluminium	3.8-4.3	3.8-4.3
copper	0.10 max	0.75-1.25
magnesium	0.03-0.06	
iron	0.10 max.	
nickel	0.006 max.	same
cadmium	0.005 max.	
tin	0.002 max.	
zinc	remainder	remainder

(The percentages in the columns from iron-tin are regarded as impurities.)

Magnesium alloys

Table 7.5: Magnesium-base alloys (BS 2970:1972)

BS designation	Mag 1	Mag 3	Mag 4	Mag 7
Percentage of				
aluminium	7.5-9.0	9.0-10.5	–	7.5-9.5
zinc	0.3-1.0	0.3-1.0	3.5-5.5	0.3-1.5
manganese	0.15-0.4	0.15-0.4	–	0.15–0.8
zirconium	–	–	0.4-1.0	–
copper	0.15 max.	0.15 max.	0.03 max.	0.35 max.
silicon	0.3 max.	0.3 max.	–	0.4 max.
iron	0.05 max.	0.05 max.	–	0.05 max.
nickel	0.01 max.	0.01 max.	0.005 max.	0.02 max.
magnesium	remainder	remainder	remainder	remainder

To a lesser extent, various of the brasses may be die cast, and for limited purposes, particularly for bearing liners, the lead-base and tin-base white metals may also be used.

Gravity Die-castings

These are produced in a metal die working under the same principle as for sand casting, where the force of gravity is relied upon to feed the metal and fill the mould.

Gravity castings are usually made in the aluminium and

magnesium alloys and, to a smaller extent, brass. Zinc-base alloys are not very satisfactory on account of the coarse grain which is promoted by this form of casting. When making gravity castings, the molten metal is poured into the die from a ladle and, in order to fill the die and compensate for the large shrinkages which take place with most of the metals, a large runner is incorporated in the construction of the die. This is afterwards cut off and used in a subsequent melt. In order to remove the casting the die must be split, and any cores necessary must be so constructed that they can be collapsed for withdrawal from the casting.

If a collapsible metal core is not practicable, a sand core is used. Fig. 7.20 shows the die for an aluminium piston with a three-part collapsible metal core for inside, the core being collapsed for withdrawal by extracting the centre wedge piece.

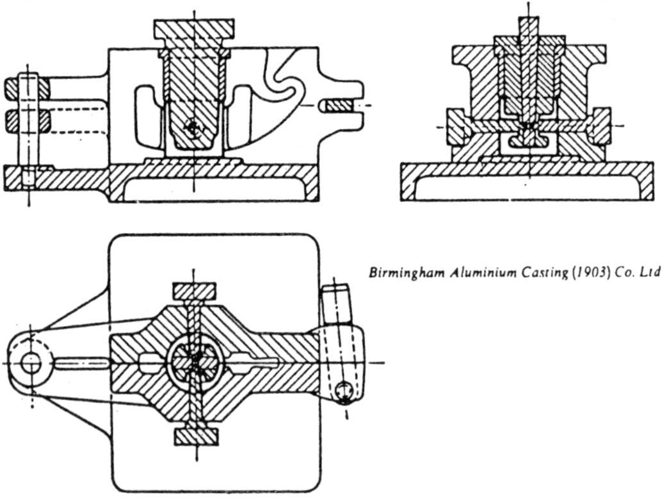

Fig. 7.20: 63 Mould for the gravity casting of a piston.

Pressure-die-casting Processes

The essentials for successful working are to ensure that the molten metal makes intimate contact with the die impression (for sharp, defined castings) and to maintain an adequate feed of metal to give sound castings. The pressures employed vary up to several hundred newton per square millimetre. There are three systems of working.

The air-blow system. In pressure castings the metal is fed to the die under pressure in some form of die-casting machine. Up to about 1934 this was the principal process in use for pressure castings. A section showing the essential details of the machine is shown in Fig. 7.14, where it will be seen that the gooseneck vessel is mounted on trunnions and immersed in a bath of the molten metal.

Tilting the gooseneck backwards, and opening the filling nozzle, permits metal to flow in, after which the vessel may be swung to the operating position for the discharging nozzle to be seated against the entrance orifice of the die. Compressed air at about 40 bars is then admitted, above the metal, foreing it at high velocity throgh the small orifice into the die.

The finished casting is afterwards extracted by opening the dies, and the long feeder, which will be cast onto it, is cut off. This method is rather restricted in the choice of alloys to which it may be applied (mostly aluminium) and has a tendency to produce porous castings.

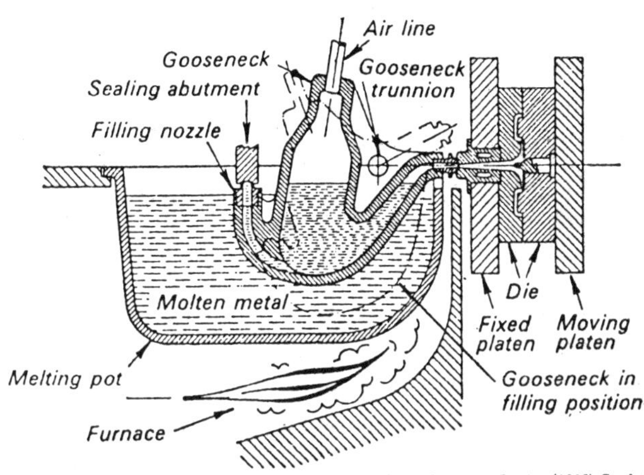

Fig. 7.21: Diagrammatic part-section through a direct air-operated gooseneck pressure-casting machine.

It has, to a certain extent, been superseded by other methods, but it is still employed for castings of the type which lend themselves to its application.

The cold-chamber system. The cold-chamber system was

first develope in the year of 1934 and consists of a cylindrical 'cold chamber', a hydraulically operated plunger, and the mould (Fig. 7.22).

In operation, molten metal from a separate furnace is poured into the chamber and then forced into the die through a suitable passage or gate. The pressure employed is extremely high (200-1400N/mm^2) and the dies have to be substantially and accurately made to avoid spring and flashing. This method is used principally for castings in the aluminium alloys and to some extent for magnesium.

Yellow metals may also be cast this way, although the necessary high temperatures and pressures limit the life of dies.

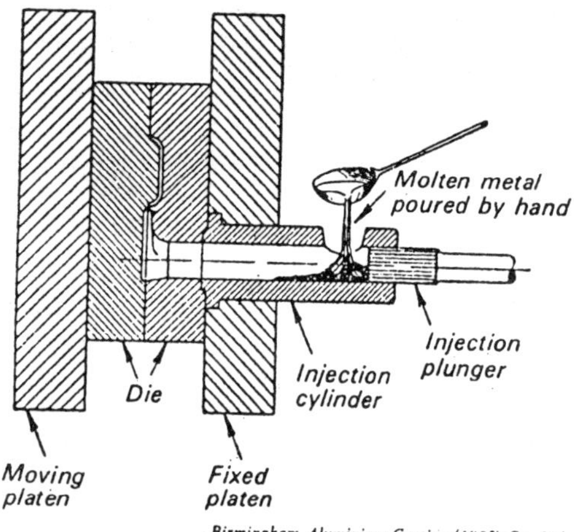

Fig. 7.22: Diagrammatic part-section through a 'cold-chamber' pressure-casting machine.

The hot-chamber system. To produce a casting, the injection plunger is operated and forces metal under a pressure of about 125N/mm^2 through an orifice into the die.

The plunger may be actuated manually through a system of levers, pneumatically or hydraulically. This system is widely used for the production of zinc-base castings and is capable of high outputs. It is not suitable for working with aluminium, but can be used for the other low-melting-point alloys such as the tin and lead groups. Machines in which the above principles are incorporated are designed with all the other facilities necessary for producing castings at the maximum rate. These include a

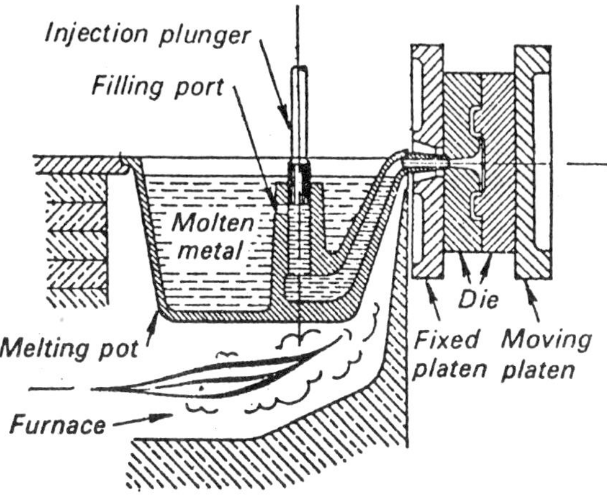

Fig. 7.23: Diagrammatic part-section through a 'hot-chamber' pressure-casting machine.

self-contained furnace, temperature control, etc., pneumatic or hydraulic means for opening and closing the dies and ejecting the finished casting, and so on.

Characteristics of die-castings

Die-casting is a highly specialised technique and, while the results often seem almost unbelievable, these are achieved only after long experience and expensive research. These castings, when properly designed and produced in first-class dies, are more than good substitutes for sand castings, and constitute an advance into a new field of production.

In the state at which they leave the dies, good pressure castings compare lavourably with many machined products, and it is common practice to put them into service without any subsequent machining. The reader might imagine that a pressure casting, produced as it is, must de denser and have superior mechanical properties to a gravity-produced casting. This, however, is not the case, because in spite of our ingenuity we cannot, in this instance, improve on what is produced by natural means.

Pressure castings can certainly be made faster, and with sharper die impressions than those fed under gravity, but the speed with which the die is filled, often requires much skill in design to avoid porosity and similar faults.

Gravity castings, therefore, are generally of more reliable structure than those made under pressure. As far as practice has developed, it is possible to produce aluminium gravity castings up to about 60 kg mass and the capacity of the pressure process is about 2000 cm^2 of projected area at the die joint, by 250 mm deep.

The capacity of the pressure process is up to about 5 kg, 3 kg and 12 kg respectively in aluminium, magnesium and zinc.

Casting features

In the following we will abbreviate aluminium, magnesium and zinc to Al, Mg and Zn, respectively.

Cored holes. Possible proportions of cored holes are

minimum: Al and Mg, 3 mm dia. x 5 mm long,

Zn, 1.5 mm dia. x 2.5 mm long;

advancing proportionately to: Al and Mg, 3 diameters long at 6 mm dia. And above,

Zn, 6 diameters, long at 6 mm and above.

Taper in cored holes: Al and Mg, 1° up to 10 mm dia., rising to 2° on diameters over 25 mm.

Zn, half the above.

Tolerances in cored holes: Al and Mg, about ± 0.04 mm on diameters up to 20 mm plus a similar amount for each additional 25 mm.

Zn, about ± 0.025 mm up to 20 mm plus 0.025

mm for each additional 25 mm.

Hole centre-distance tolerances. Under the most favourable conditions tolerances of ± 0.05 mm up to 40 mm centres, and ± 0.025mm for each additional 25 mm may be obtained. When cores are located in different members of the die it might be necessary to add up to ± 0.08 mm to the above.

Spigots, tenons, etc. These may be cast to fit without subsequent machining.

Wall thickness. On the average run of castings, the wall thicknesses recommended are as under:

Gravity-cast Al and Mg: minimum 3 mm. No maximum provided metal can be fed from an adequate riser.

Pressure-cast Al and Mg: minimum 2 mm, maximum 5 mm

" " Zn " 1.5 mm " 5 mm

Small areas allow thinner sections under favourable conditions, e.g. areas of 5-10 cm^2 may be cast in thickneses ranging from 0.5 to 0.75 mm. Tolerance on wall thickness about ± 0.08 mm.

The above tolerances refer to pressure castings; for gravity work they should be at least doubled.

Cast inserts. The process lends itself to the casting in of components made of another metal, and examples of these (see Fig.7.24), when it will be seen that various methods are adopted for providing anchorage.

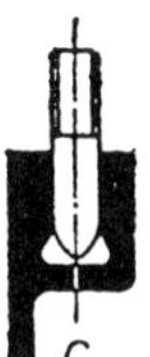

Fig. 7.24: 67 Die-cast inserts

Screwthreads and gears. For most commercial purposes threads may be pressure cast in zinc alloy, and external threads

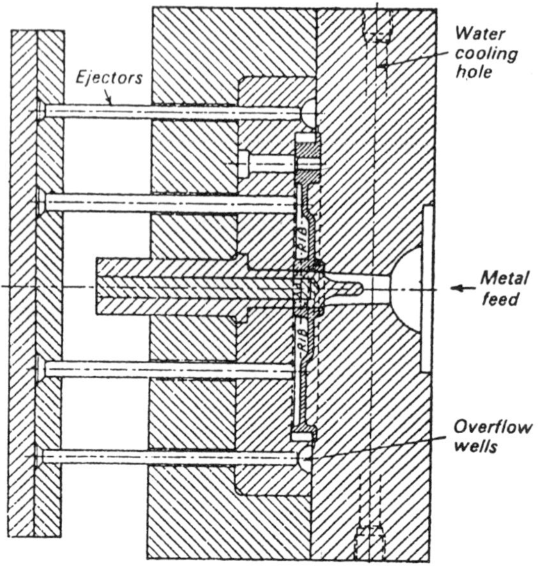

Fig. 7.27: Section of mould for pressure casting a gear (approximately 140 mm diameter).

in Al and Mg. The thread is formed by a screw or ring incorporated in the die. If the member inserted for casting the thread is solid, either the casting must be screwed out of the die or the screwed mould (detachable) screwed from the casting after ejection.

A split die or ring may be employed but the joint line may leave a flash necessitating a finishing operation.

The threads should be as coarse as possible, but pitches up to 1 mm may be cast in Zn, and to 1.5 mm in Al and Mg. Plain spur and bevel gears and chain wheels may be produced in all the alloys, and the method has the advantage that the steel shaft or bush for the gear, provided with suitable arrangement for anchoring the casting, may be inserted in the mould for the gear to be cast onto it.

A complicated mould for pressure die casting is shown in Fig. 7.25 which shows a section of the dies for casting a gear with shrouded teeth. The metal is fed through the right-hand platen, and the long pins carried in the left-hand plate are for ejecting the finished casting from its impression.

8

Sheet Metal Work

When engineering goods and households are produced, sheet metal work is used, where other methods are unviable or too complicated. A brief description of the trade is given below to provide the reader a good working knowledge of the subject concerned.

This is an art of drafting pattern from working drawings. It must be realised that the main duties of the draughtman is to give dimensions and shape of the finished job. They are least concerned with surface developmentof the various parts except in estimating the approximate sizes of the metal required. Sheet metal worker must be familiar with hand tools, laying out of patterns, methods of forming joints and cutting plates to exact size.

VARIOUS HAND TOOLS

Following are the sheet metal working hand tools:

1. Different types of hammers
2. Mallets
3. Stakes or forming supports
4. Shears or snipes.

Stakes

This is usually used as supporting and forming tools. They

also help in bending operations. For performing different operations they are fitted on the bench and then the job is worked on them (Fig. 8.1). A *half moon stake* is used for working the edges on discs.

A *hatchet stake* is used widely for forming, bending and seaming the edges. *Bick iron* is used for forming long tapered cylindrical components. For conical components a *funnel stake* is used.

A *horse head stake* is used for bending, holding and supporting the components.

A *convex stake* is used for forming spherical shapes. A pipe stake is most suitable for forming pipes and hollow cylindrical surfaces.

Shears

Punch shears are used in sheet metal works are made in different shapes to suit various requirements like cutting grooves, holes, seams and edges. A hollow punch is used for producing small circular holes.

A grooving punch is used for making grooves. In addition to these tools various types of chisels used in sheet metal working are flat, round, half round and knife edge.

The straight shear is used for cutting along a straight line whereas bent shear is used for cutting along a curvature. Hand shears are very light and can be used for cutting with one hand but they cannot be used for cutting sheets above 0.8 mm thickness.

Marking tools. In sheet metal work like the various measuring tools used scale, height gauge, try square, wing compass, dividers, straight edge, vernier calipers and micrometer. For details of these instruments refer to chapters on

(i) marking and measuring instruments, and

(ii) bench work and fitting.

Sheet Metal Joints

There are several sheet metal joints which are used in carpentry as follows:

(i) Lap joint,

(ii) Seam joint,

(iii) Locked seam joint,

(iv) Wired edge joint,

(v) Hem joint,

(vi) Cap joint,

(vii) Flanged joint, and

(viii) Angular joint.

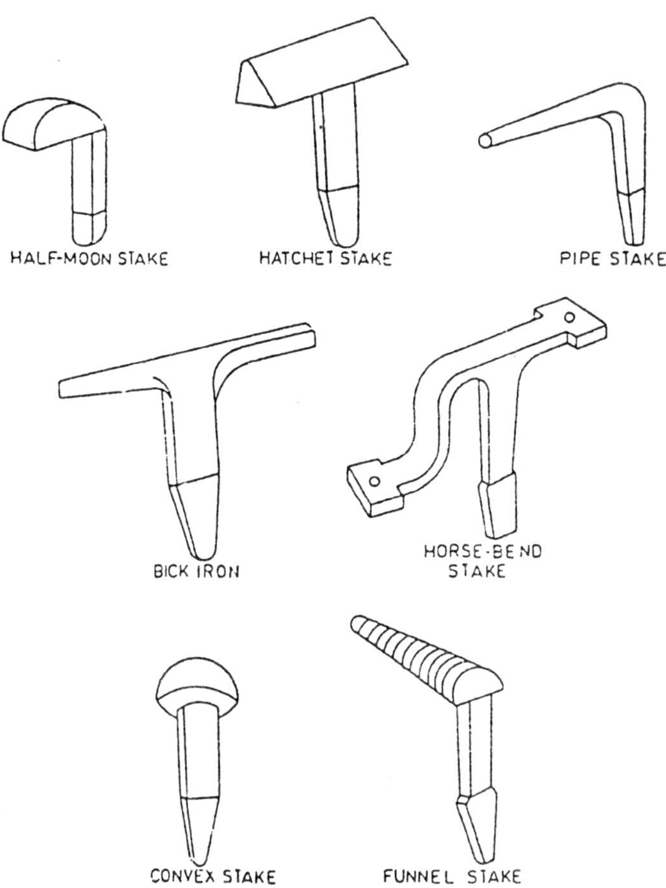

Fig. 8.1: Stakes used in sheet metal work.

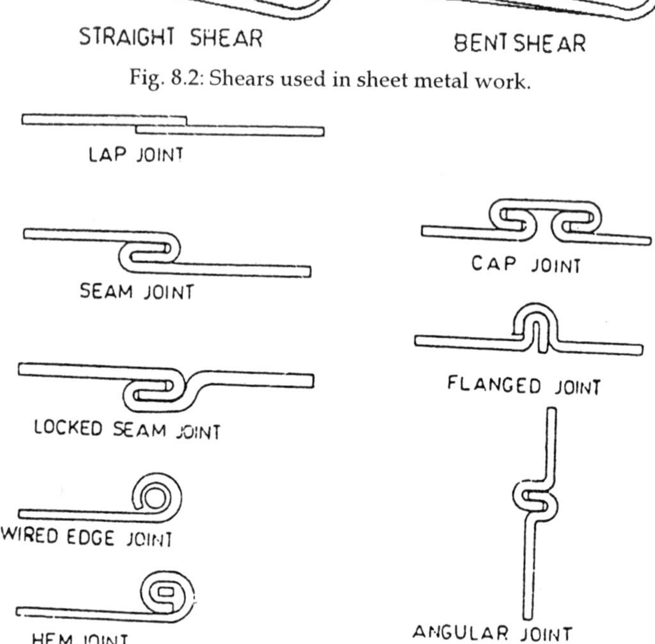

Fig. 8.2: Shears used in sheet metal work.

Fig. 8.3: Various types of joints used in sheet metal work.

DEVELOPMENT OF THE LAYOUT

Layout development of a component in sheet metal work may be said to be the brain that controls the actual process of production. It is very difficult to laydown a set of standard rule for pattern development of a job. The procedure used for development of components with square or rectangular bases is given below:

(a) Draw the elevation of the component in a right position.

(b) Draw a vertical line and mark a convenient point on it.

(c) From this point measure down the top ordinates or half-widths.

(d) Number the points.

(e) From the same point measure down the half-width of the base and draw a line of indefinite length from the point.

(f) Step off various lengths along the horizontal line from the corner of the angle formed.

(g) Join the points marked in the correct order.

(h) Cut the lengths and join them to make the component.

CONE PATTERN

When a cone shown in Fig. 8.4 is developed, this pattern is called cone pattern. Draw the elevation of the cone and produce the edge lines to the apex O. For the pattern mark the point 0 in a suitable position. With O as centre and OB as radicus draw an arc of indefinite length. Measure the circumference of the cone, preferably with a flexible steel rule. Draw an arc with radius *OA*. Cut *BD* equal to the circumference at *BD*. Develop the cone from the pattern.

The second method of development of the frustum of the cone is shown in Fig. 8.5. Draw a semicircle on the bigger diameter of the cone as shown. Mark 0,1, 2, 3, ...16 on the sheet. Select a suitable point *O* and with *O* as centre and *OA* as radius draw an arc of indefinite length. Mark 16 points (number can be taken as per convenience) on the arc and complete the pattern. Draw deep lines with a scriber, cut the sheet and form the cone with a suitable joint.

TAPERED LOBSTER-BACK BEND

A tapered lobster-back bend shown in Fig. 8.6 and the method for development of a pattern. It consists of three segments tapering gradually from a large to a smaller diameter pipe. The joints between the segments are either welded or butted together. For joining, it is necessary to make metal thickness allowances for a slip in the joints, such that the sheet sets over each other. Marks 1 to 6 shows the diameter of the large pipe and 7 to 13 the diameter of the smaller end of the same pipe.

Similarly the reduced diameter of this pipe acts as the

major diameter of the other pipe. The reduced diameter of the second pipe is equal to the major diameter of the third pipe.

There are a number of segments of the required size. The pattern for each segment is drawn out separately. Consider the development of the half pattern of base segment.

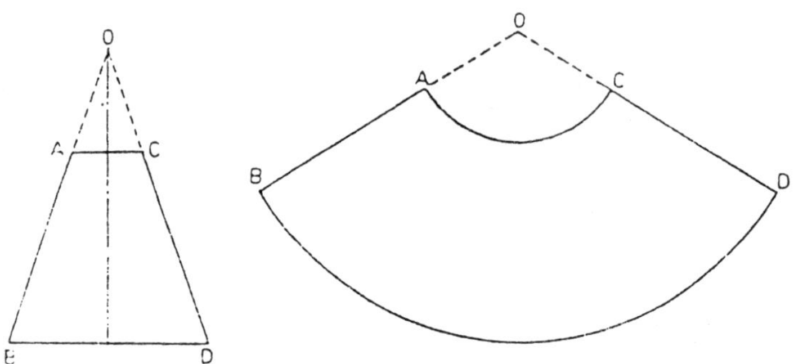

Fig. 8.4: Development of pattern of the frustum of a cone.

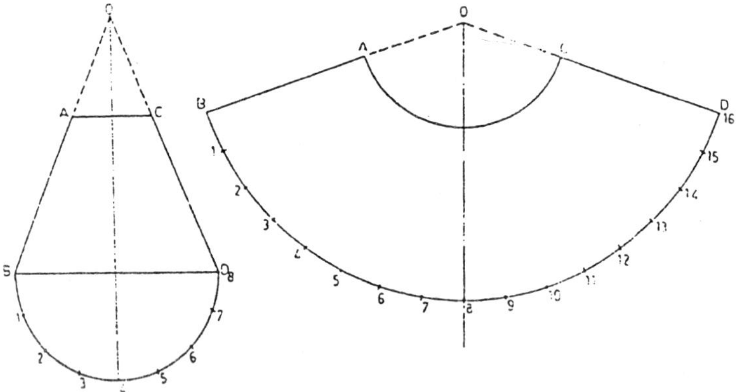

Fig. 8.5: Development of pattern of the frustum of a cone.

(a) Divide the segment into a convenient number of triangles (six shown in Fig. 8.6). Draw a semicircle on A-6 to represent half the circumference of the large end of the segment. Divide the semicircle into six or eight equal parts and draw perpendiculars to the base line A-6. Next draw a semicircle on 7-13 and draw perpendiculars to the top edge line. Join all the points with false length lines to obtain elevation triangles.

Complete the arcs with a free hand and mark dark lines on it. Cut the pattern and develop the cone. Develop the patterns similarly for circular segments (b) and (c). Finally join these segments to form a tapered lobster-back bend.

METAL THICKNESS

Allowances for metal thickness need consideration unless the metal sheet used is very thin. This can be well understood from the following practical example:

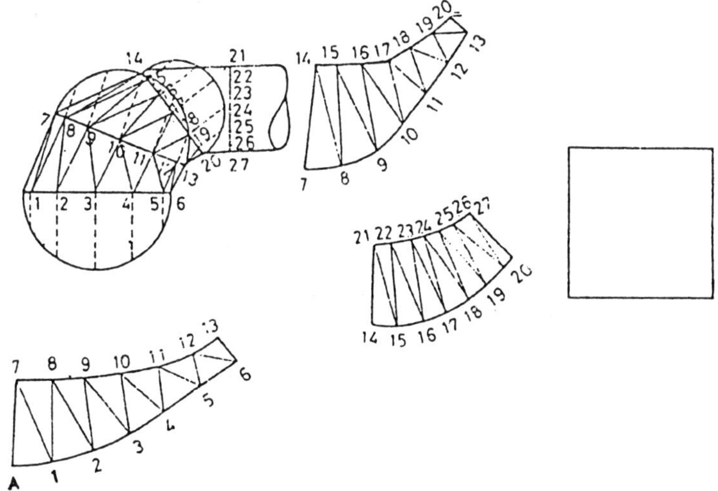

Fig. 8.6: Development of pattern of tapered lobster-back bend.

Suppose there is 9 cylinder of 15mm mean radius as shown in Fig. 8.11. Let the thickness of the metallic sheet be 1 mm. Thus the outer diameter of the cylinder is 16 mm and inner diameter is 14 mm. The length of dotted line is 15 x 3.1416 = 47.124 mm. The outer length of the sheet is 16 x 3.1416 = 50. 265 mm while the inner length is 14 x 3.1416 = 43.9824 mm. Thus, if the outer diameter of the cylinder is given, then we have to take the mean diameter for calculation of size of the sheet. The correct length of sheets (outside diameter—metal thickness) is 3.1416.

In the same manner allowances are needed for square rectangular and bent surfaces. Standard bend allowances used on sheets are given in following table:

Table 7.1: Bend Allowance Table

Swag No.	Bend allowance	
	inches	millimetres
10 (.128 in)	.502	12.25
12 (.104 in)	.408	10.25
14 (.080 in)	.314	7.85
16 (.064 in)	.251	6.30
18 (.048 in)	.188	4.75
22 (.028 in)	.110	2.76
24 (.022 in)	.850	2.15

WORKING MACHINES OF SHEET METAL

To increase the rate of production a large number of power operated machines have been developed. The various machines used in sheet metal working are:

1. Shearing machine,
2. Folding machine,
3. Grooving machine,
4. Swaging machine,
5. Beading machine,
6. Peining machine,
7. Bending machine,
8. Beam bending machine, and
9. Straightening machine.

In addition to the above-mentioned machinery many special purpose machines are manufactured to perform selected operations particularly for mass production of identical parts.

Shearing machines: These are used in sheet metal working can be divided into the following two categories:

1. Hand operated shearing machine
2. Mechanical shearing machine

Above then 0.8 mm thickness, power operated machines

are used for cutting sheets up to 6 mm thickness. Both the machines consist of two high carbon steel/alloy steel hardended blades. One blade is fixed while the other blade is movable. A hand operated machine is fitted on the table while power operated machine is robust in construction and grouted on the floor. see Fig. 8.12.

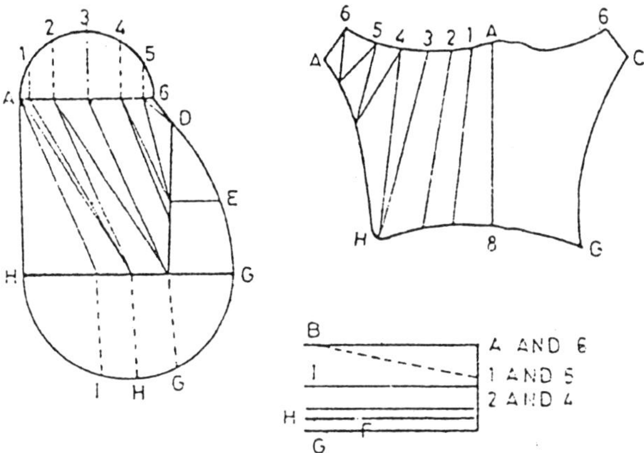

Fig. 8.7: Development of pattern of breaches piece.

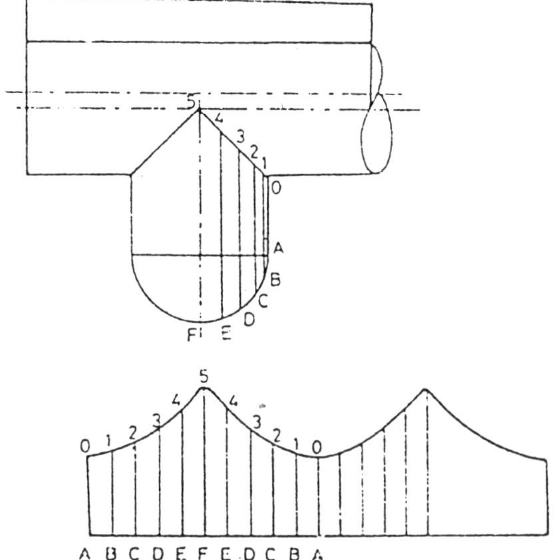

Fig. 8.8: Pattern for T-pipe pieces of equal diameters.

Folding machines. Folding are used for folding the sheet at the edges to form the joint.

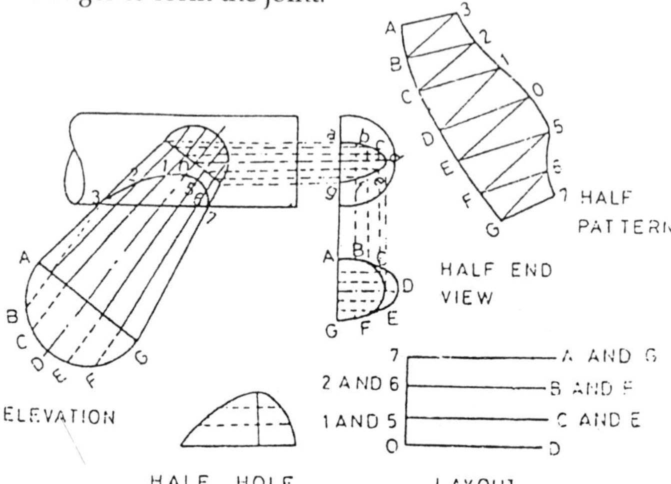

Fig. 8.9: Pattern for conical hopper on inclined plane.

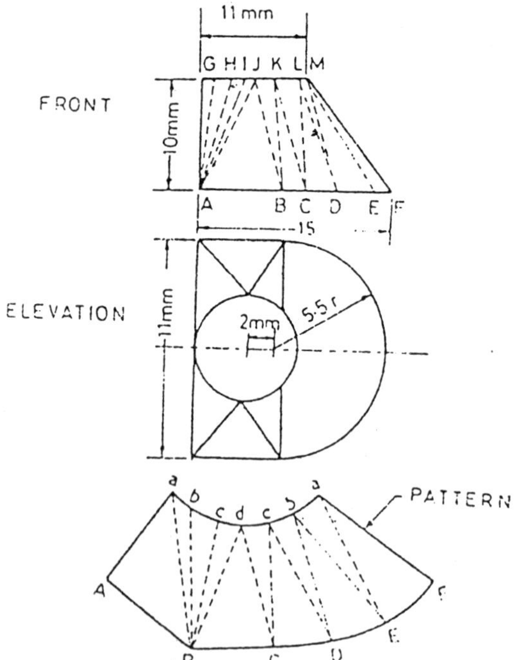

Fig. 8.10: Pattern for round hopper with flat back.

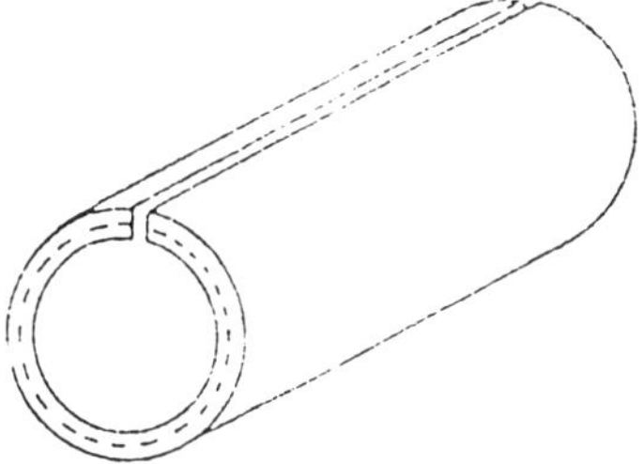

Fig. 8.11: Metal thickness allowances.

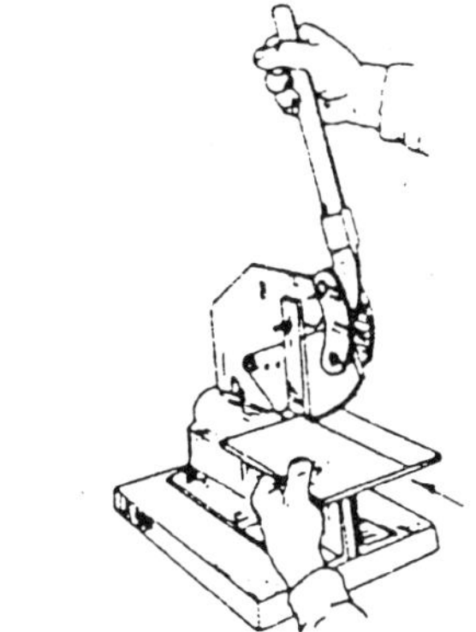

Fig. 8.12: Mechanical shearing press.

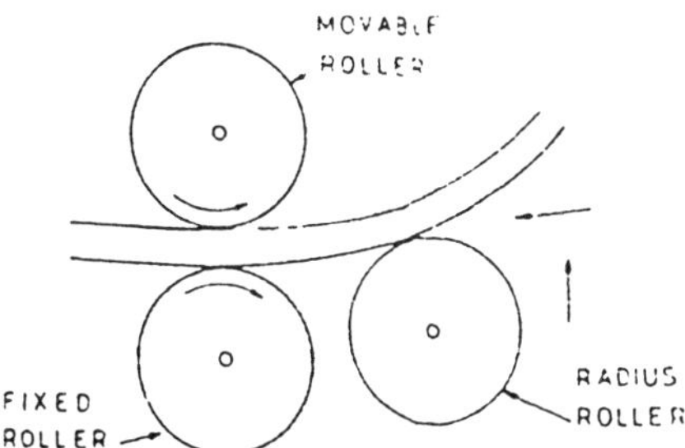

Fig. 8.13 (a): Principal of sheet bending.

A *grooving machine* is used for forming grooves in sheets. *Swaging machines* are used to provide different types of contours on sheets.

These machines are used for shaping metal sheets into cylindrical objects. The machine consists of three rollers that can be adjusted for different radii. In different types of machines these rollers have different adjustments.

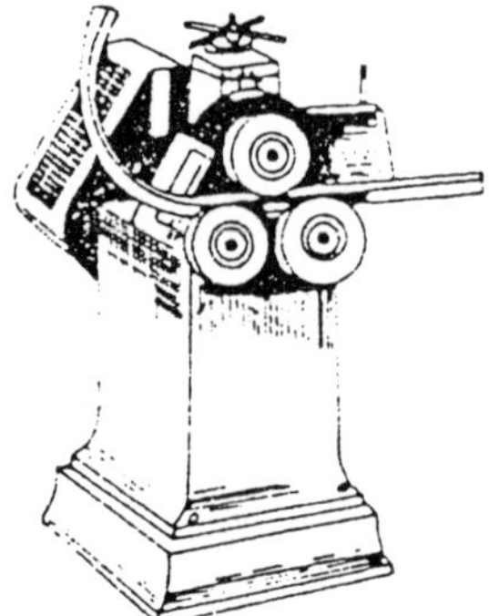

Fig. 8.13 (b): Sheet bending machine.

All the rollers can be moved vertically up and down to suit the desired positions and provide different curvatures. But the most suitable control is shown in Fig. 8.13 (a).

Improved designs of these machines use rollers of different grooves to provide different shapes on sheets.